GIS *for* Water Management

IN EUROPE

Mike Bedford

ESRI Press
REDLANDS, CALIFORNIA

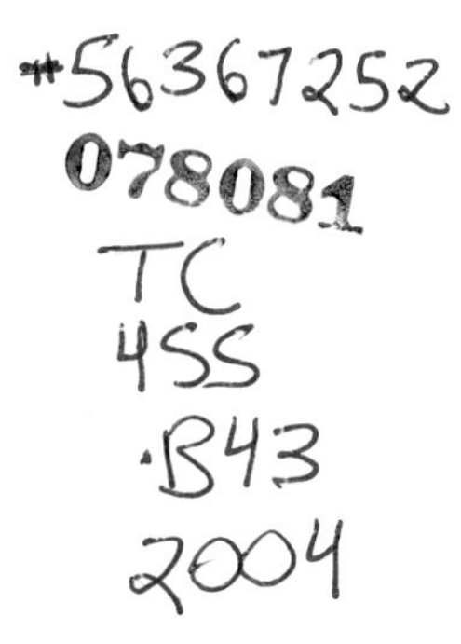

ESRI Press, 380 New York Street, Redlands, California 92373-8100

10 09 08 07 06 05 04   1 2 3 4 5 6 7 8 9 10

Printed in the United States of America

Library of Congress Cataloging-in-Publication Data
Bedford, Mike, 1954–
GIS for water management in Europe / Mike Bedford.
p. cm.
ISBN 1-58948-076-7 (pbk. : alk. paper)
1. Water resources development–Europe–Data processing. 2. Water-supply–Europe–Management–Data processing. 3. Water quality management–Europe–Data processing. 4. Flood damage prevention–Europe–Data processing. 5. Geographic information systems–Europe. I. Title.
TC455.B43 2004
333.91'0094—dc22
2004019779

Books from ESRI Press are available to resellers worldwide through Independent Publishers Group (IPG). For information on volume discounts, or to place an order, call IPG at 1-800-888-4741 in the United States, or at 312-337-0747 outside the United States.

ESRI Press agent for the United Kingdom, Europe, and the Middle East:
Transatlantic Publishers Group
Telephone: 44 20 8849 8013   Fax: 44 20 8848 5556
transatlantic.publishers@regusnet.com

Contents

Foreword

Water is a basic commodity, supporting life in all its forms. As a consequence of constantly increasing human pressures such as population growth, agricultural intensification, industrialisation and recreational activities, the quantity and quality of earth's water resources are more and more under threat.

In Europe, the intense use of and increasing pressures on water put it at risk of overexploitation and pollution; as a consequence, water needs careful management and protection. That fact is reflected in the number of regulations and legislation at all levels of administration and policy. The adoption in 2000 of the European Water Framework Directive (WFD) established an overall legislative framework with the scope to significantly improve the quality of all European waters within the next decade. The WFD asks for the establishment of River Basin Management Plans throughout Europe; these cross national boundaries, and they analyse pressures and impacts, and evaluate alternative management options where necessary.

Water-related natural hazards, such as droughts and floods add to the existing problems, causing serious damage to property and threatening many lives, even in highly developed countries. Increasing evidence of climate change and an expected increase in the frequency of extreme events call for improved management from local to European scales, a fact that is increasingly acknowledged at the political level.

In the industrial societies of Europe the management of water involves the design, implementation and monitoring of highly complex systems, ranging from water distribution and sewage systems in communes to the management of water resources across extended, international river basins such as the Rhine and Danube. The management of water resources through numerous administrative and technical bodies, as well as the frequent crossing of national borders, requires the harmonisation of data across administrative and political entities. In parallel, Europeans will need to set up intelligent data management systems that can connect, retrieve and analyse information from different databases maintained by local authorities. The thematic questions to be solved are manifold and depend on the scale of the issues at stake and the types of problem to be studied.

A common aspect of all these applications, however, is the spatial dimension of the phenomena and the dynamic changes to be modeled. This is the reason why GIS plays such a prominent role in the field. GIS technology can help oversee the spatial relationships in complex systems, to locate problems and to plan and evaluate alternatives. It can help to identify critical areas and pathways and to model the flow of water, nutrients, pollutants and sediments through the hydrological system (including surface runoff, soil infiltration, groundwater bodies, lakes, rivers and the sea).

This book presents a collection of examples of how GIS technology can help to solve a variety of problems related to the management of water resources in Europe. Among others, these include issues such as the harmonization of spatial information along administrative borders, the search for environmentally sound options to build new waterways, the analysis of environmental pressures and the establishment of early warning systems for floods and avalanches.

Given the political and technical evolution in the field of water management and the rapidly growing capabilities of modern spatial data handling and analysis, this book presents timely illustrations of current options to solve real-world problems, which were designed, tested and implemented in the real world.

This book is not a GIS manual, nor a GIS textbook. Through the presentation of eleven examples, Mike Bedford (with contributions from Maarten Vergouwen) illustrates the use of GIS technology in solving water management problems and in providing water services to citizens. Without indulging in excessive technical detail, the book illustrates the widespread use of GIS in this field. Even though most cases here have been implemented using ESRI® technology, the book's emphasis is not on specific software, but rather on the basic philosophy and solutions behind spatial data handling and analysis in the water management sector.

It is my hope that this book will help to create greater awareness of the role and the possibilities of GIS in the field of water management and its impact on our daily life—that it helps to inform the reader of the need for long-term and sustainable management of this precious resource.

Ispra, Italy,
June 2004

Jürgen Vogt
European Commission
DG Joint Research Centre
Institute for Environment and Sustainability

Acknowledgments

This is not the sort of book that could have been written in isolation. By its very nature, a book of case studies requires input from a large number of people and I'd like to express my thanks to them here. Thanks to the staff at the various European ESRI offices and distributors for suggesting suitable stories and making the initial contact with the clients. Thanks also to those people with interesting stories to tell who provided me with information, photographs and screen shots, who answered my questions, and who checked the final copy. Their names appear at the end of each chapter.

In fact, work on three chapters was taken almost entirely out of my hands. Thanks to Maarten Vergouwen for his work on chapters 2, 8 and 11, which relate to water management in the Netherlands. Dorothy Wiers provided able and speedy translation help for those chapters when she was called upon, so many thanks are also due her.

Normally I spend my time writing magazine articles. Pulling together a complete book has been a new and rewarding experience for me and I'd like to thank Christian Harder at ESRI Press for giving me the opportunity. Also, at ESRI Press, I'd like to express my gratitude to Richard Greene, my editor.

Mike Bedford
Keighley, United Kingdom

1

Navigating the Danube: Joining East to West

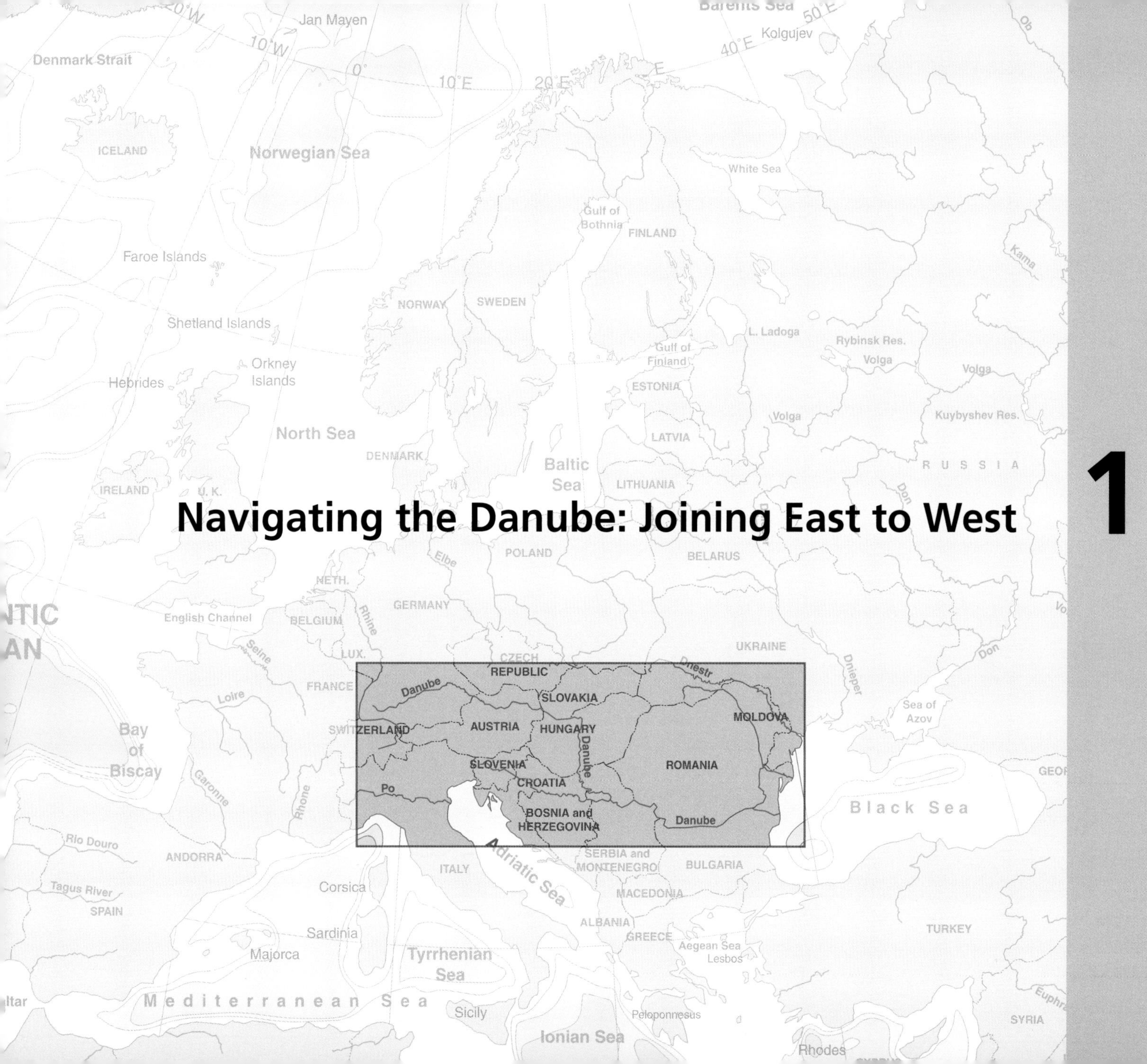

At 1770 kilometres, the River Danube is Europe's second longest, passing through five major cities and ten countries. The completion in 1992 of the Rhine-Main–Danube canal created an uninterrupted waterway from Rotterdam to the Black Sea—a 3500-kilometre backbone for the European economy.

Low water levels in this section of the Danube in Bavaria often impeded the passage of vessels, creating a severe bottleneck in one of Europe's most important economic arteries.

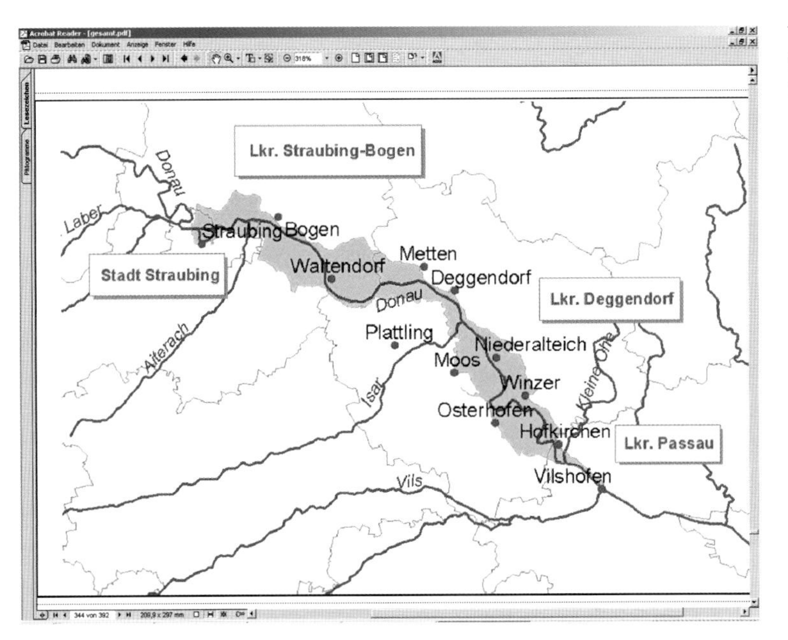

The highlighted portion of the map shows the floodplain of the Danube, extending through three German Landkreise, or counties.

However, one 70-kilometre section of the river in Southern Bavaria, between Straubing and Vilshofen, is a bottleneck in this otherwise wide-open artery, because water levels there are frequently too low to allow ships to pass. Throughout the 1990s, environmental studies assessed the types of engineering work that would be required to open up that section of the river to year-round navigation. Several plans were devised, and five were selected for further study. Recommendations included construction of locks, barrage dams, bypass canals and other structures—or various combinations of all of these—to free ships' passage down one of the greatest rivers in the world.

GIS formed the basis for the winning plan. When work based on this plan is complete, sometime before 2018, Europe will benefit from a much improved transportation system, while environmental risks will be kept to a minimum.

Issue-based GIS

To assess the impact of the various river modification proposals, an issue-based GIS, called Do-GIS—named after the river's German name, Donau—was created. Do-GIS combined all relevant data regarding the river and its floodplain, the technical construction variables, the affected land use, and landscape scenery. Do-GIS has been used to evaluate and assess all this information using a flexible methodology that combines data analysis, modeling and visualisation technologies.

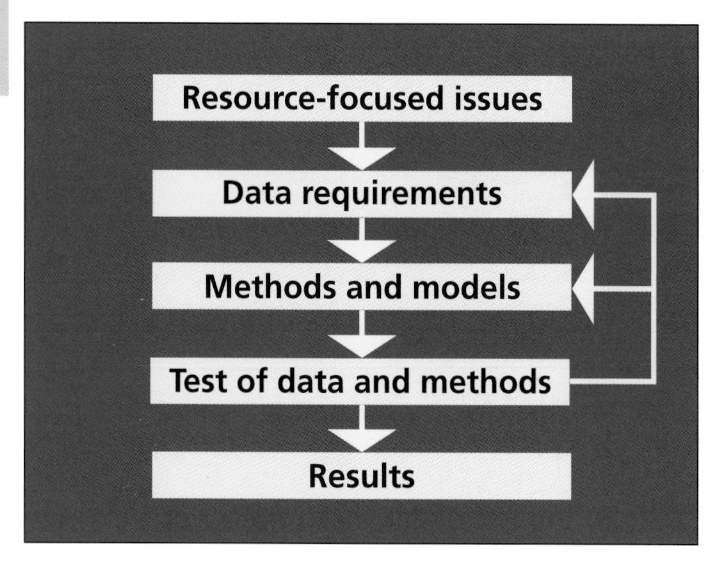

Schematics of the organisational structure of Do-GIS

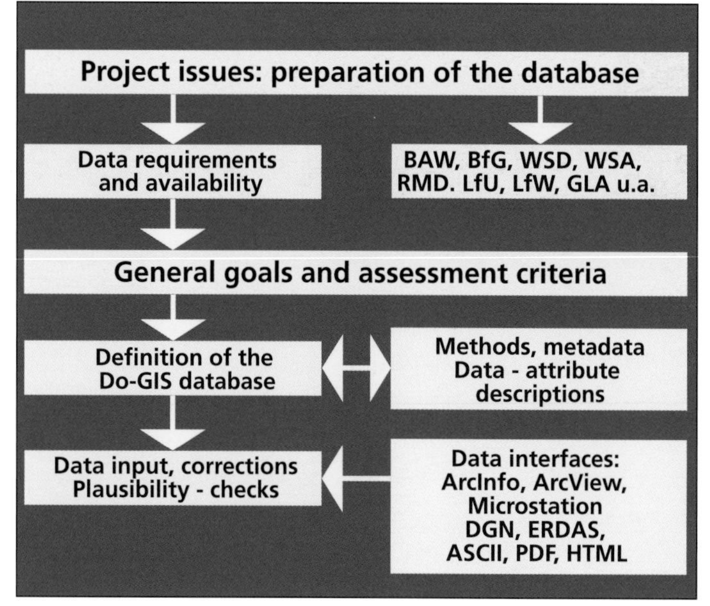

Topographical and photogrammetrical information in the Do-GIS database included aerial photography of the study area.

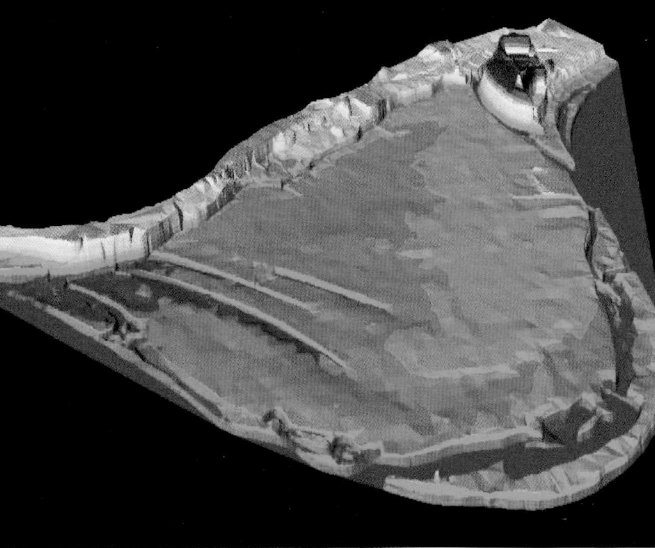

Photography of the area was supplemented with maps and a three-dimensional digital terrain model.

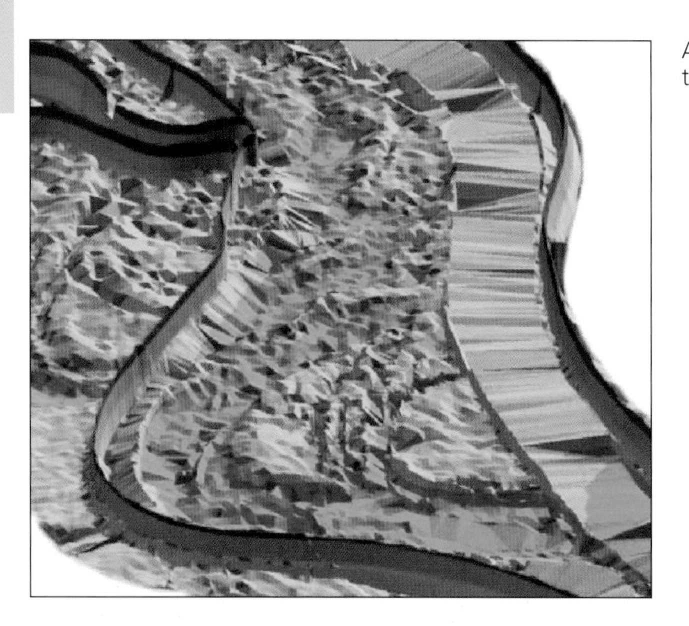

A sonar-derived model of the riverbed also formed an essential part of the topographical element of the database.

Do-GIS will help answer three central questions:

- What are the actual ecological and land-use conditions in the floodplain?
- How will these conditions be changed by the construction of each of the different engineering alternatives?
- How can the various proposals be best explained and visualised for the public and political decision makers?

These questions in turn influenced the system architecture, the database design and selection of the GIS, and the approaches to models.

Database architecture

To organize the answers to these questions, the Do-GIS database was structured into four broad themes encompassing a wide range of data. This included topographic and photogrammetric basemap information such as topographic maps (1:25 000 and 1:5000 scale); digital terrain model; digital riverbed model (sonar data combined with DGPS); water level data derived from photogrammetric interpretation; Digital Federal Waterway Map of the Danube River (DBWK); aerial photographs; digital ortho images; and color infrared and other imagery.

The four database themes encompassed abiotic and biotic data. Abiotic data included data sets of hydrology, morphology, hydrogeology and soil conditions; hydrology of the Danube; morphology of the Danube; material and nutrient exchange cycles; hydrogeology (groundwater conditions); and soil conditions in the floodplain.

Biotic data included information about vegetation, species and plant communities, breeding birds and guest birds in the floodplain and on the river, amphibians, fish species (including those in effluent rivers and creeks), macrozoobenthos (invertebrate animals larger than 2 millimetres) and molluscs in the Danube and in the floodplain.

In addition, human-use requirements were collected, including data for recreation, landscape scenery and quality, land use, agriculture, forestry settlements, water consumption and management, nature protection and cultural values.

Data sets were compiled from various federal and state agencies. Many were collected during past studies, while others were newly created for the ecological study. Some already were in ArcInfo® coverage format, while many others came from CAD systems or as ASCII output from numerical models.

Numerous data sets on abiotic, biotic and human aspects were already available. All of them had been mapped during past studies, at a scale of 1:5000. For each theme, 81 map sheets had to be merged for the ecological study.

In addition, CAD data from the Digital Map of Federal Waterways and from construction plans was converted to coverage format. Scanned topographical maps served as basemaps.

Evaluation and model application

Once the Do-GIS database was complete, it was used to compare the environmental impact of the various proposed modifications with the status quo. This evaluation process involved coupling Do-GIS to hydrological and groundwater models to determine the result—on the river and its floodplain—of the engineering options. In particular, the

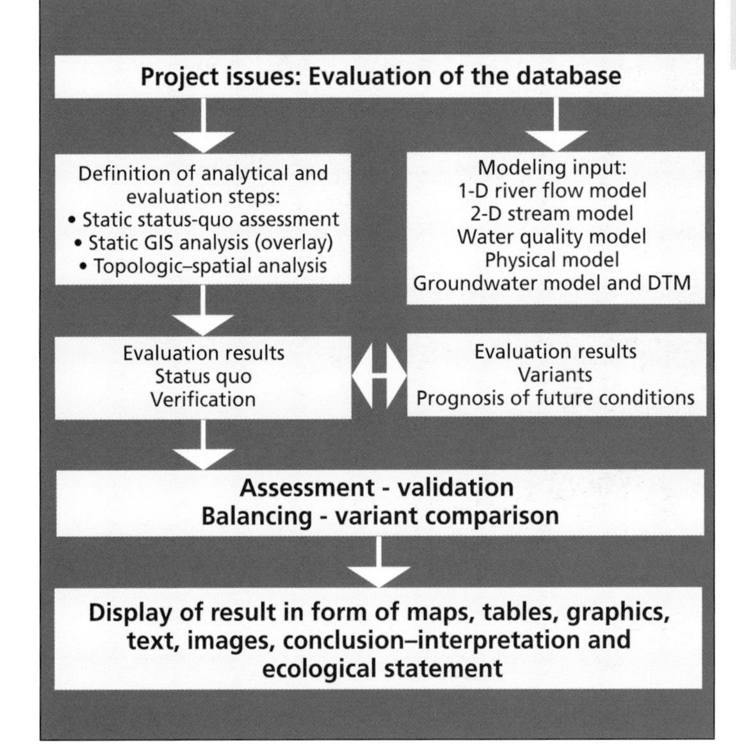

For each technical variant, Do-GIS assesses the environmental impact, presenting the results as maps and as three-dimensional visualisations.

output from the model, comprising surface water quality, flood potential, and groundwater levels and dynamics, was analysed to determine the impact on the various ecosystems. The results were presented spatially as shown in the maps reproduced on this page and the next. This GIS-based approach allowed experts to evaluate the ecological impact according to previously determined criteria and thereby select those options that most closely matched the optimum.

1

1. A triangulated irregular network (TIN) was used to create a digital terrain model of the floodplain, accurate to 100 millimetres. 2. The groundwater levels were then calculated using the groundwater model. 3. Next, a vegetation map was generated. 4. The impact of groundwater levels on the vegetation was then assessed. 5. The layer of vegetation sensitivity was combined with the layer of groundwater levels to show the necessary balance that must be achieved.

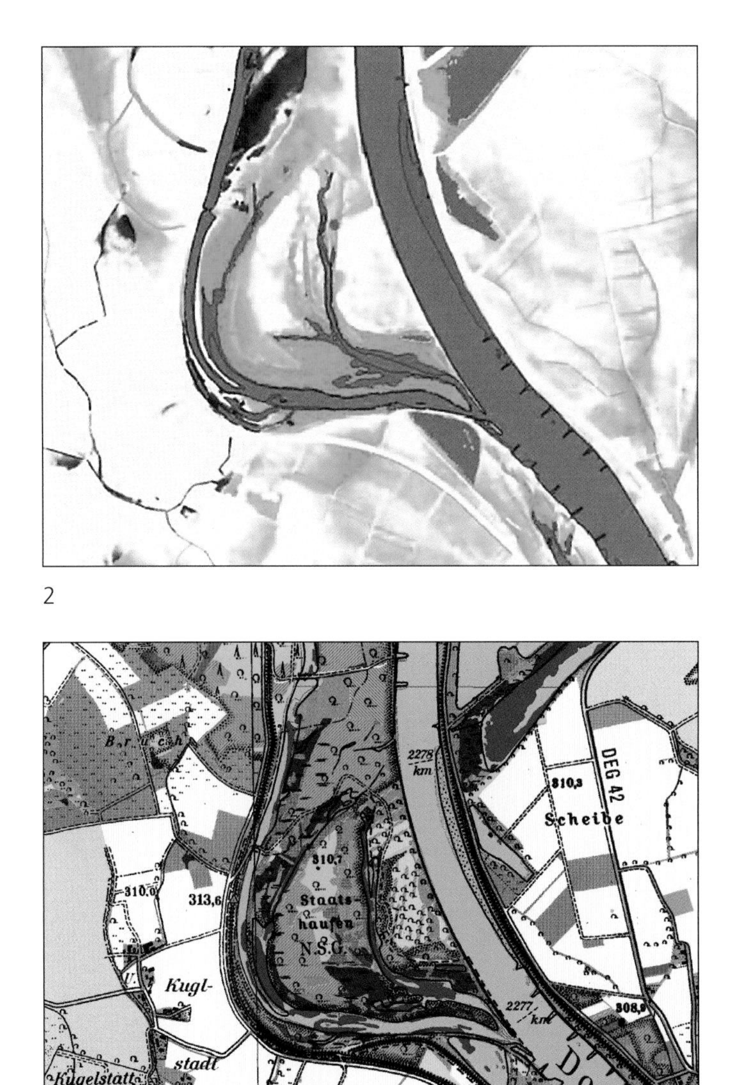

DEG 42
310,3
Scheibe
2278 km
310,7
Staats-
haufen
N.S.G.
310,0
313,6
Kugl-
stadt
Kugelstatt
2277 km
308,9
309,3
311,0
Donau

2

3

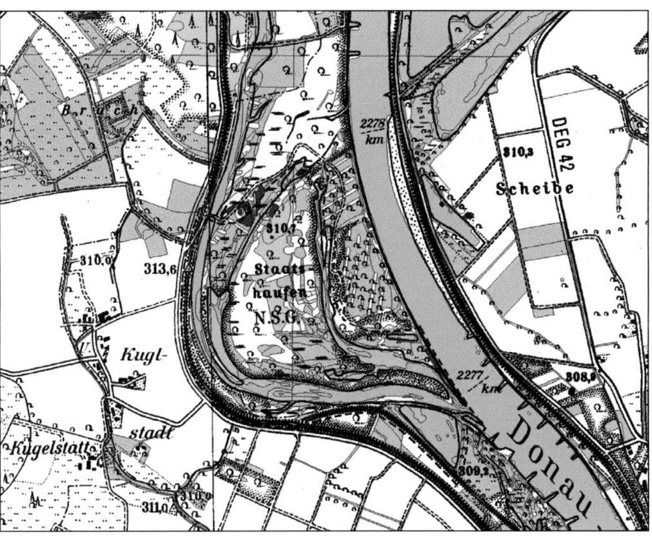

2278 km
DEG 42
310,3
Scheibe
310,7
Staats-
haufen
N.S.G.
310,0
313,6
Kugl-
stadt
Kugelstatt
2277 km
308,9
309,3
311,0
Donau

4

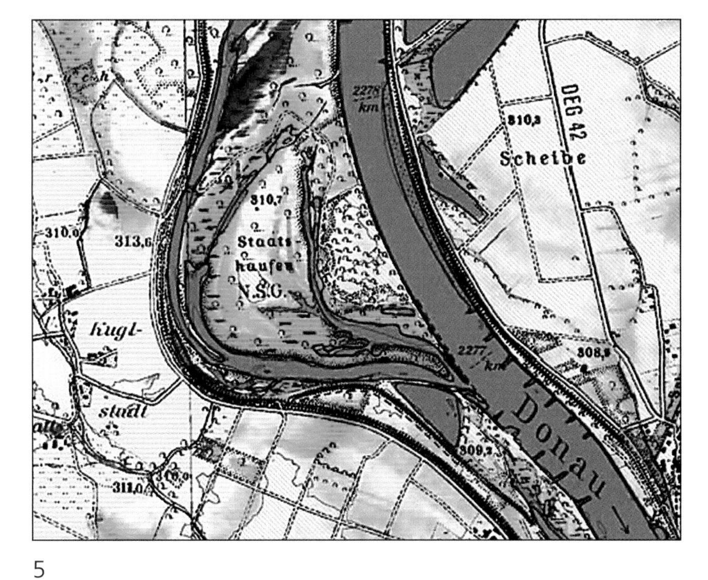

2278 km
DEG 42
310,3
Scheibe
310,7
Staats-
haufen
N.S.G.
310,0
313,6
Kugl-
stadt
2277 km
308,9
309,3
311,0
Donau

5

Three-dimensional visualisation of the impact on the landscape scenery

It was important to be able to visualise the impact of the various construction plans on the landscape scenery and compare them with current conditions. This was achieved using an advanced 3-D GIS application. Input to that application included data on the existing landscape as well as information from Do-GIS on how the appearance would change as a result of the construction. Expected changes were due to the construction measures themselves, the new land-use management, and changes to vegetation expected to result from the new ecological conditions.

The 3-D models were based on digital terrain models and aerial photographs of the current landscape. For each construction variant this information was integrated with CAD data, model output on the vegetation changes due to ecological changes, and new land-use management plans. The images shown here are visualisations of the actual situation and the new landscape scenarios that take into account the vegetation development and the new land use.

Hardware
Compaq® Proliant servers

Software
Microsoft® Windows NT® Server 4, ArcInfo Workstation 7.2, ArcView® 3.1 and 3.2, ArcView Spatial Analyst 2, ArcView 3D Analyst™, ArcGIS® 8.1, ERDAS IMAGINE® 8.4, World Construction Set 4 PC, Adobe® Acrobat® 5

Data
All data sets are entered on a standard form where the source and mapping method are documented, as are details of document processing.

All spatial data in Do-GIS is in coverages, grids, TINs, shapefiles and georeferenced TIFF image formats. Attribute data is kept separately in tabular format (DBF, XLS), but can be linked back to the corresponding geometries using unique identifiers. Metadata is maintained in Microsoft Word and published as Adobe Acrobat PDFs.

Acknowledgments
Dr. Fritz Kohman and Dr. Michael Schleuter from the German Federal Institute of Hydrology, Koblenz

Prof. Joerg Schaller, ESRI Germany

Klaus Rachl, Wolfgang Steib and Thomas Hack of Planingsburo Prof. Joerg Schaller, Kranzberg

Berhard Holfter, Informationssysteme, Leipzig

RMD Waterway Company, Munich

2

Friesland: Keeping the sea tamed

Water dominates the landscape and life of Friesland, a pancake-flat province in the north of the Netherlands: the North Sea stretches along an extensive coastline on one side, while inland, myriad ditches, canals, rivers and lakes predominate. In summer Friesland is a haven for water sports enthusiasts, but its principal economic business is agriculture, with most of the land used for grazing by herds of distinctive black-and-white Frisian cattle.

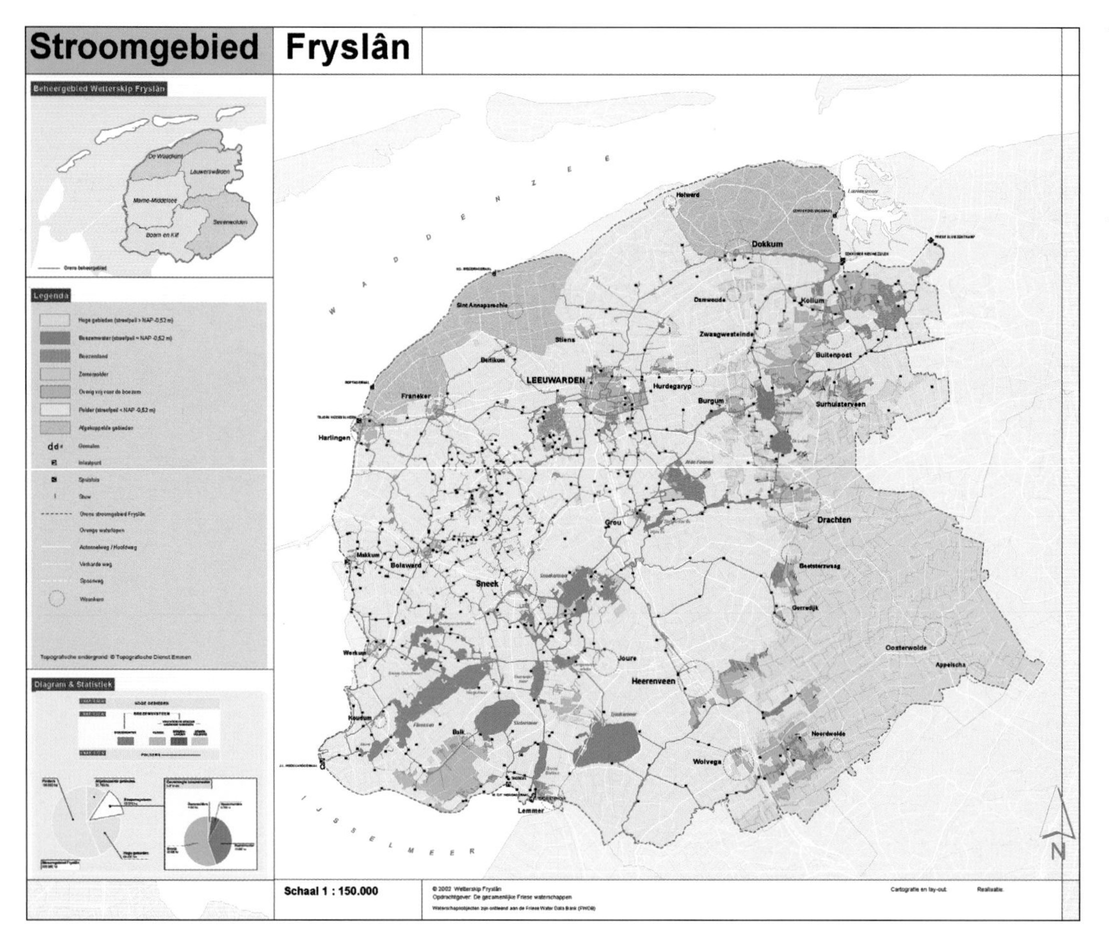

This map of the management area of the Frisian Water Board shows both the extent of the board's area of responsibility and the extent of the province's problems. Boezem water is indicated by the blue areas, while boezem land is indicated by dark green; polders and areas available for boezem use are indicated by the other shades of green; high land (areas higher than 0,52 metres below sea level) is designated by orange.

In order to sustain its economic well-being, much energy in Friesland is devoted to keeping the sea tamed. Two-thirds of the Frisian drainage basin, which is 335 000 hectares in size, is below sea level. This area consists largely of polders, which are areas of land reclaimed from the sea, kept dry by an intricate network of dykes, ditches, dams, pumping engines and boezems or drainage outlets. To keep the polders dry—and the important agricultural areas—is an unceasing, technically complex operation.

Currently, the intricate Frisian water management system functions well, which means the economy functions well, and Frisians can go about their daily lives. But radical changes in the climate of northern Europe are predicted for the coming decades and that means that additional flood-prevention measures will most certainly be needed for the future. GIS technology helped Friesland's water managers plan for these measures.

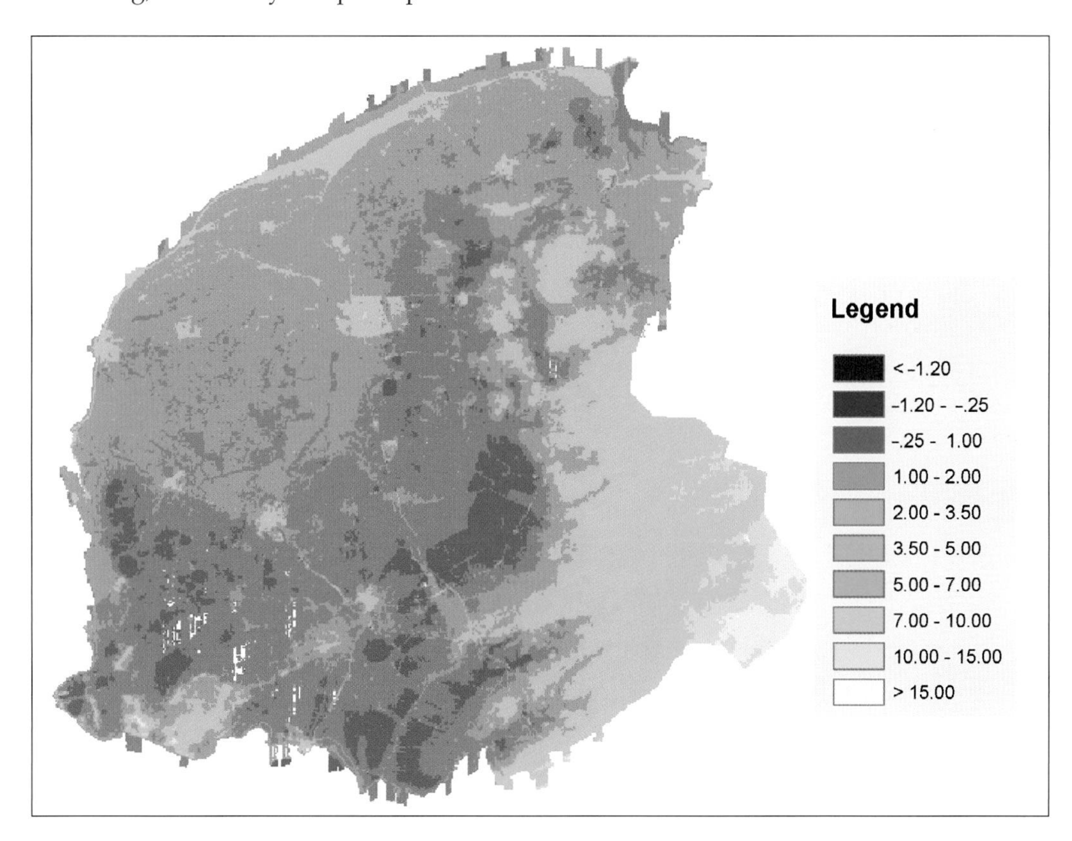

Friesland's land elevation in metres in relationship to sea level, with darker blue indicating areas below sea level, and greens and yellows, above sea level.

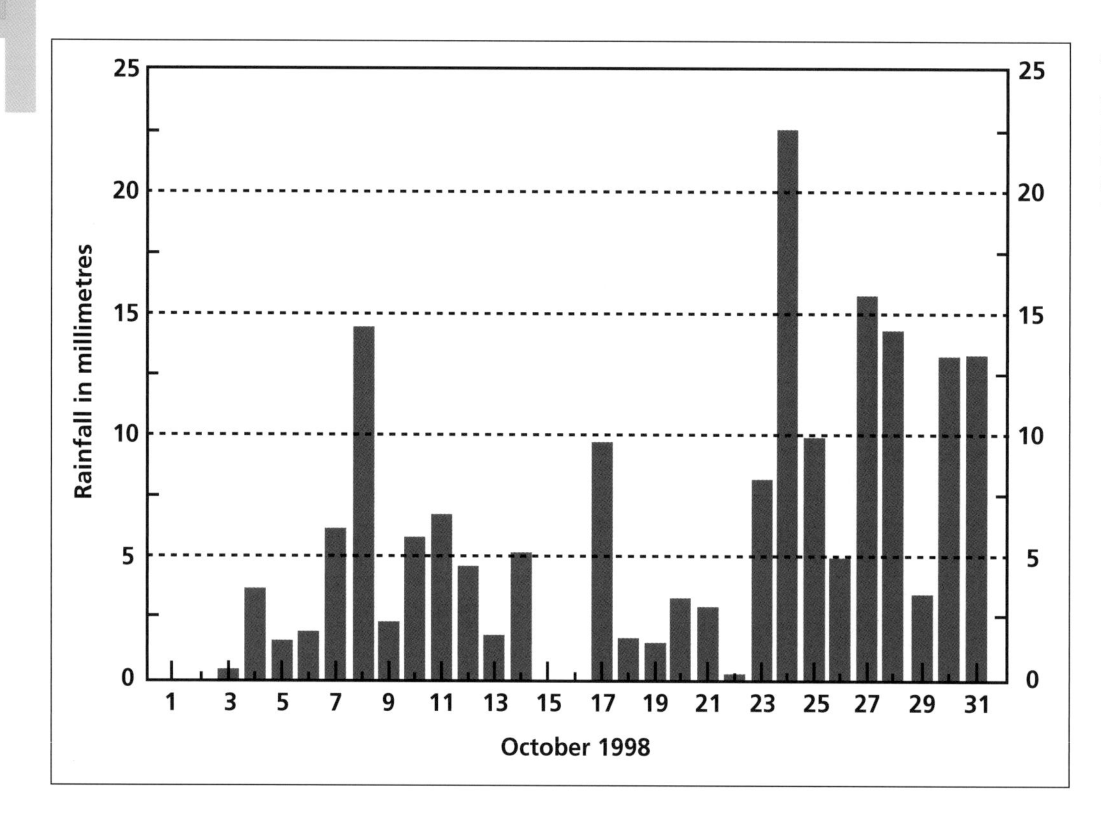

A daily rainfall chart for October 1998, one of the wettest Octobers in recent Friesland history, and a possible harbinger of things to come. Total rainfall for the months was 179 millimetres; normal is 72 millimetres.

Grim forecast

Meteorologists predict that not only will more rain fall in the Netherlands in the coming years, the rainstorms will be heavier. By 2100 a 20-percent increase in precipitation is expected. Climatologists predict that simultaneously, the mean sea level around the Netherlands could rise by as much as 600 millimetres. This combination of factors means that, almost certainly, flooding danger in Friesland will increase as well. To make matters worse, the extraction of natural gas and salt in the north of the region, together with an increased settling of peat and clay, are expected to cause a further drop in the ground level. These factors only serve to increase the risks.

The biggest problem of a polder is getting the water out of it. The usual procedure is to drain the water away through a series of ditches to a pumping engine, which pumps it over an inland dyke and into a boezem. When the level of water in the boezem rises above a certain level, it is in turn pumped out by a larger pumping engine, across another, higher dyke, and into a river, canal, lake or the sea. But large rainstorms can produce so much excess water that the main pumping engine that should empty the boezem cannot cope.

When extreme rainfall occurs at a time when strong northwesterly winds are blowing, the situation becomes even worse. Such winds along the Waddenzee cause the water level along the coast to rise, also increasing water levels in the polders and overloading the drainage system even more, to a point where pumps could fail completely.

A number of extremely wet periods in 1995 and 1998 showed just what might happen in the future. All over Friesland water levels rose, almost to the top of the inland dykes; on several occasions, it was touch and go whether the boezem barriers would hold. Flooding would mean a rise in water level of between one and two metres—unlikely to result in catastrophic loss of life, but causing enormous economic damage nonetheless.

Fortunately, the water receded, and significant flood damage was averted. This close call made it obvious, however, that some additional measures were required to lessen future flood risk.

Drainage basin surface area (in hectares)		
Higher Friesland ground	64 030	
Boezem system	42 310	
Boezem water		15 060
Boezem lands		27 250
Polders	195 880	
Low-level land that remains undrained (e.g., nature preserves, marshlands)	31 760	
Total area	**333 980**	**42 310**

The components of the Friesland drainage basin: more than half of the total area is made up of polders. The 42 310 hectares of boezem system are used to help regulate the polder areas.

Additional measures

Deciding what those measures would be was the task of the provincial government and the regional water board, in consultation with other interested parties. Among the options to be considered: widening the ditches, enlarging the boezems, stricter controls in environmental planning, building extra pumping stations and, in extreme conditions, banning or limiting drainage of the polders.

Because of the complexity of the decisions and the potential controversy surrounding them, many parties—water managers, local and provincial authorities, conservationists, farmers, barge owners operating on the inland waterways,

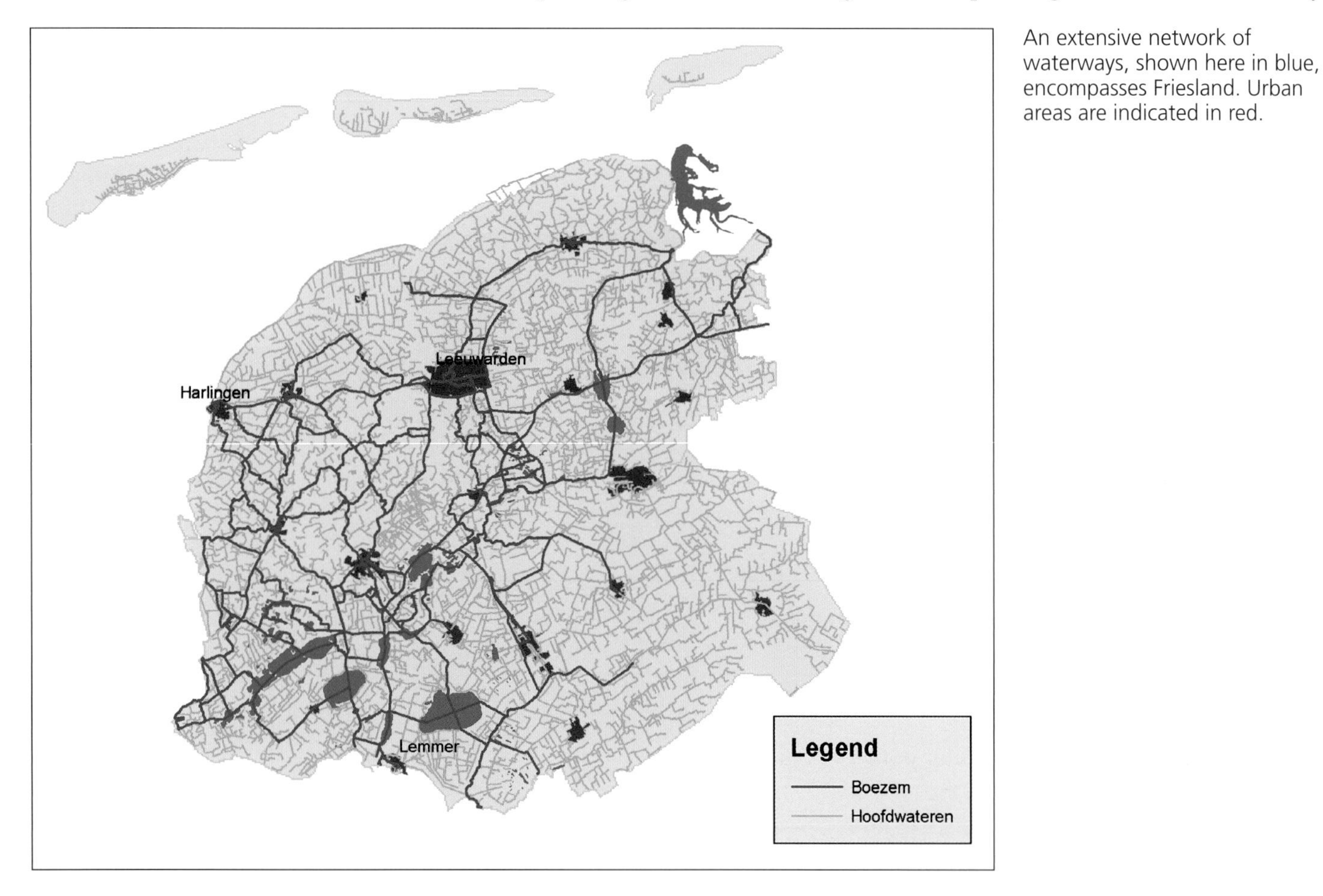

An extensive network of waterways, shown here in blue, encompasses Friesland. Urban areas are indicated in red.

local residents, people involved in the tourist industry and others with their own unique interests in the water system—have all been watching the debate with great interest.

Limiting drainage to the polders, for example, is a controversial step for farmers. In this scenario a choice is made to stop draining water from a particular polder to relieve pressure on the boezem. This would lead to a higher level of water in the polder, and so, damage to crops and grassland.

GIS technology has been used to clarify these multiple and often conflicting interests by delivering maps that show exactly where water becomes a problem in times of bad weather, how often such flooding can happen, and the potential damage from that flooding.

Charts for decision making

Friesland's water board must decide what protection is needed, and the degree and nature of damage that might be acceptable in a community. As part of this calculation, the water board has adopted flooding safety standards. These are determined using statistical models to gauge the frequency with which extreme meteorological conditions might occur, usually at intervals of 10, 100 or 1000 years. By combining this information with detailed local geographic information—such as the height of a particular dyke—the water board created inundation charts. These show what part of the land will flood in the event of a 10-year, 100-year or 1000-year flood event. Because climatic predictions show that the situation is certain to worsen, the charts have been compiled with three different starting dates—now, and the years 2030 and 2100, to reflect the climatic conditions that are predicted to exist in those years.

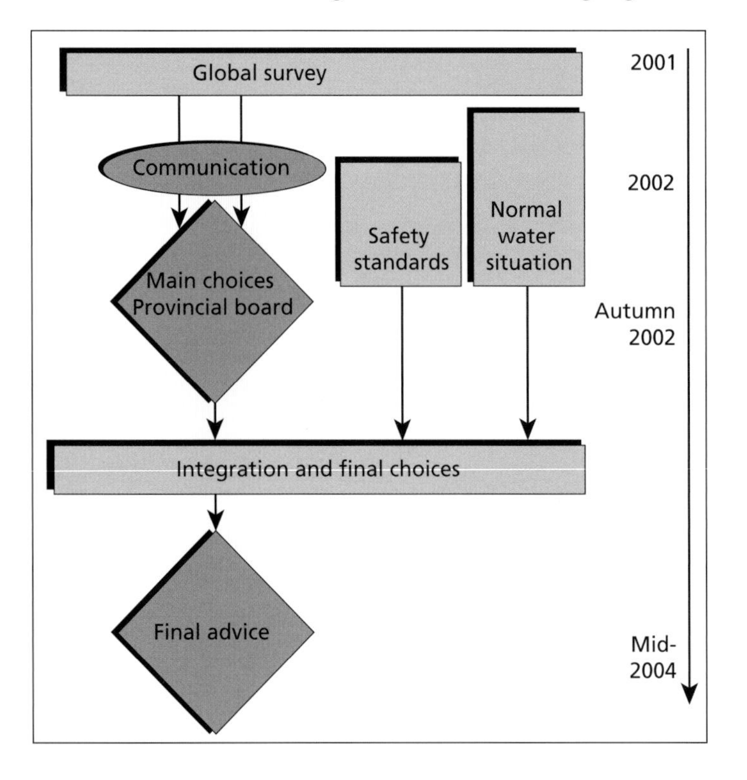

Acceptable inundation levels vary according to land use: protecting an urban area is considered more critical than protecting farmland, because of the larger number of lives that would be disrupted. To clarify this process, water risk charts have been compiled on top of the inundation charts.

The charts clearly indicate whether the land within an area is adequately protected against flooding. The charts show three types of land use: arable land, grazing land and urban development. Flood risk in these areas is further divided into three types. Areas coloured red do not meet the flooding safety standard. Orange-coloured areas are designated for a particular land use, but from a water management point of view the risks are too great and planning permission should be withheld. Green areas indicate that an area meets the flooding safety standard.

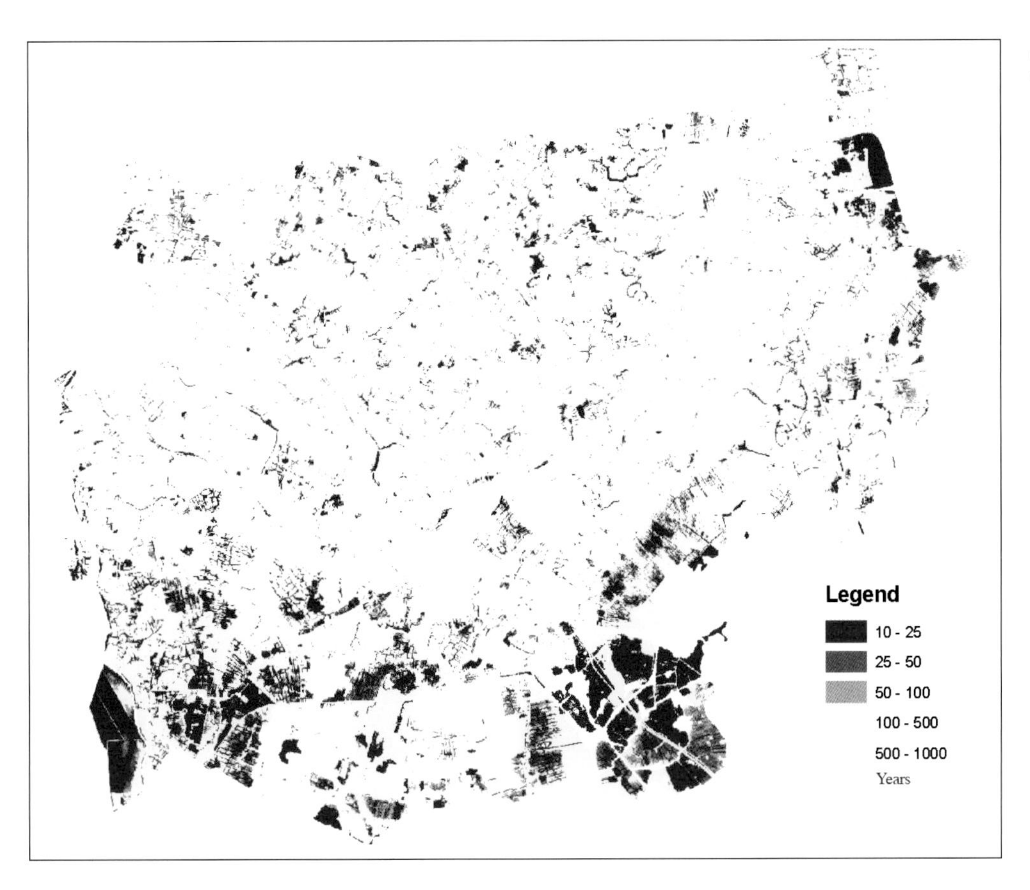

Areas in red will flood most likely once every 10 to 25 years.

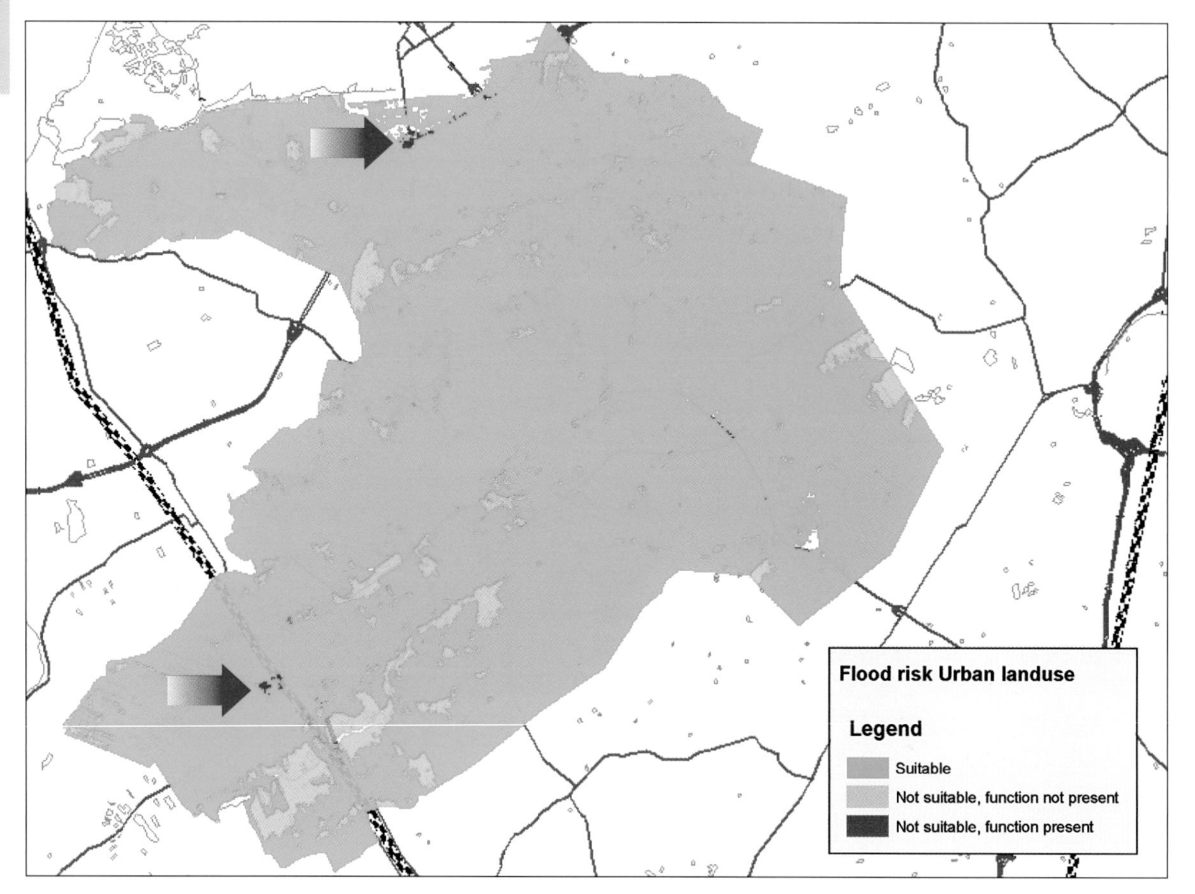

Flood risk is combined with an urban land-use map, which in turn can be used to estimate overall economic damage from a flood event.

SOBEK and GIS

An important facet of the study was to discover exactly how much information is required to determine flooding risks, and thus, whether the charts and GIS-driven maps are adequately communicating that risk; that is, the methodology itself had to be tested. So it was decided to set up a trial in which precise calculations were made for 16 selected areas of Friesland that comprise 2 percent of the entire surface area of the province. The areas represented a cross-section of the entire management area. A large amount of accurate data about water management in those areas already existed.

A water risk chart was then designed for each of the 16 areas with the use of a very detailed model of the rainfall run-off process. This model—called SOBEK, after the ancient Egyptian river god who took the form of a crocodile—produced very detailed and accurate calculations. Although too time-consuming and far too costly to apply to the entire province, SOBEK, designed by a consortium of Dutch water organizations and technical institutes, was nonetheless an effective tool for assessing the accuracy and cost-effectiveness of the simpler GIS-based models.

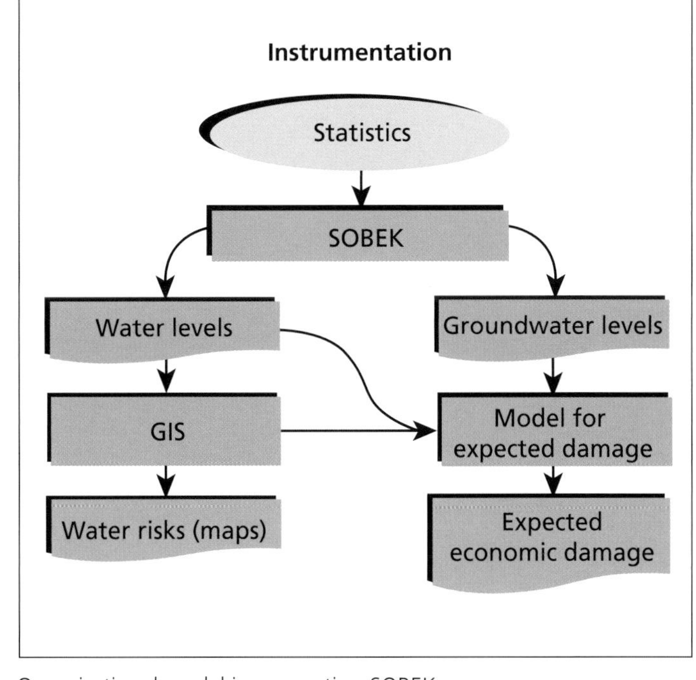

Organisational model incorporating SOBEK

Comparisons and recommendations

Water risk charts for the 16 areas were then designed, using GIS models, and the results were compared in detail with those of SOBEK to see how much they differed. These differences were plotted on water risk deviation charts according to a colour code.

The result: the differences were minimal. So it was decided to opt for the cheaper and simpler GIS water risk calculations for the whole of Friesland.

As a result of the study the project team has been able to make recommendations concerning:

- Current protection of the land against flooding
- Risks of building in the flood-sensitive areas of the province, enabling control in environmental planning
- Future measures needed to maintain the desired protection level
- Development of an emergency plan in the event of imminent danger

Software

Client: ArcGIS 8.2, Oracle® 8.1.7

Server: ArcSDE® 8.2

Data

Intwis—central information system for water boards

Acknowledgments

Thanks to Jan Heida of Wetterskip Fryslân, and to Dorothy Weirs, for translation assistance.

3

Czech Republic: Preparing for EU membership

In May 2004, 10 countries on the European continent were slated to become full members of the European Union. The event was noteworthy not only because it was the largest single expansion of the EU, but also because it included eight countries previously concealed behind the Iron Curtain.

For these new EU member countries, the EU expansion required them to meet the *Copenhagen criteria*, a set of standards to bring their government operations and administrative systems into compliance with EU standards. For many of the *acceding countries*, as they were known, doing so meant making major changes to their systems of taxation, customs, corporate law, industrial standards, transport, energy, agriculture, social policy, consumer protection and environmental protection.

One acceding country, the Czech Republic, used GIS to plan for the changes it had to make to meet the water-related environmental requirements of the EU's Water Framework Directive (WFD)—specifically, those of the Urban Waste Water Directive and the Directive for Quality of Fresh Water.

In the language of the WFD, its purpose—and thus, the mandate to the Czech Republic—was to establish a framework for the protection of inland surface waters, transitional waters, coastal waters and groundwater, a framework that would, in its words:

- prevent further deterioration, and protect and enhance the status of aquatic ecosystems, terrestrial ecosystems and wetlands directly depending on the aquatic ecosystems;
- promote sustainable water use based on long-term protection of available water resources;
- enhance protection and improvement of the aquatic environment through specific measures for the progressive reduction of discharges, emissions and losses of priority substances and their cessation or phasing-out;
- ensure the progressive reduction of pollution of groundwater and prevent further pollution;
- contribute to mitigating the effects of floods and droughts.

Like previous European water legislation, some of these aims are specific to particular categories of water. However, the emphasis of the new WFD, which came into force in December 2000, is quite different. The old legislation concentrated on regulating discharges and placed constraints only on the water quality in designated areas. The WFD tackles water quality at the river basin level and requires that all water bodies meet certain environmental standards. Like all existing EU nations, the Czech Republic was required to bring into force the laws, regulations and administrative provisions necessary to comply with this directive by December 2003.

The Czech perspective

Water management planning in the Czech Republic is carried out by Water Boards under the umbrella of the Ministry of Agriculture. Although their overall mandate was broad, an important part of their planning process was to formulate ways to meet the WFD standards. To do so, they created a Decision Support System (DSS) with which to assess policies and to select the most cost-effective way of meeting WFD requirements. The DSS includes data and modeling tools to

- provide an overview of pollution sources, river systems, water quality conditions, existing water supply and wastewater treatment facilities, as well as technical options for making improvements and tools for calculating associated costs;
- assess the changes to water quality that would result from implementing the various strategies, and estimate the corresponding investment as well as operations and maintenance costs;
- identify the lowest-cost strategies for meeting the legal requirements of the directives for water supply and wastewater treatment;
- estimate the economic and financial implications of accession to the EU, including the effect on investment programmes, recurrent costs and financing options.

The DSS provided access to various databases and modeling tools through a GIS interface, which in turn offered a user-friendly method of specifying various scenarios

'With regard to pollution prevention and control, Community water policy should be based on a combined approach using control of pollution at source through the setting of emission limit values and of environmental quality standards', states the EU's Water Framework Directive.

and an easy way of retrieving the results. The user could explore the range of options offered by different interpretations of the ultimate EU directives, as well as alternative technical strategies, specification of intermediate compliance targets and final time frame for meeting the requirements. An optimisation tool was used to identify the least costly strategies for meeting specified ambient water quality objectives.

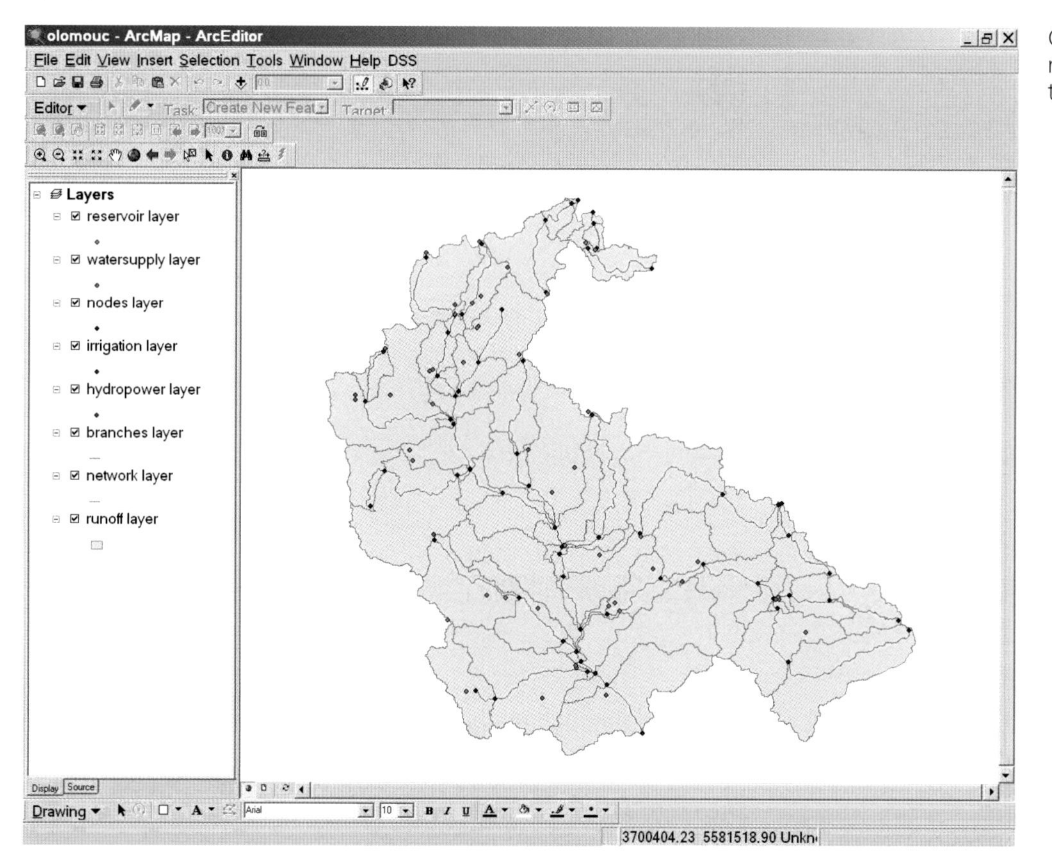

GIS facilitates access and use of relevant environmental information in the environmental planning process.

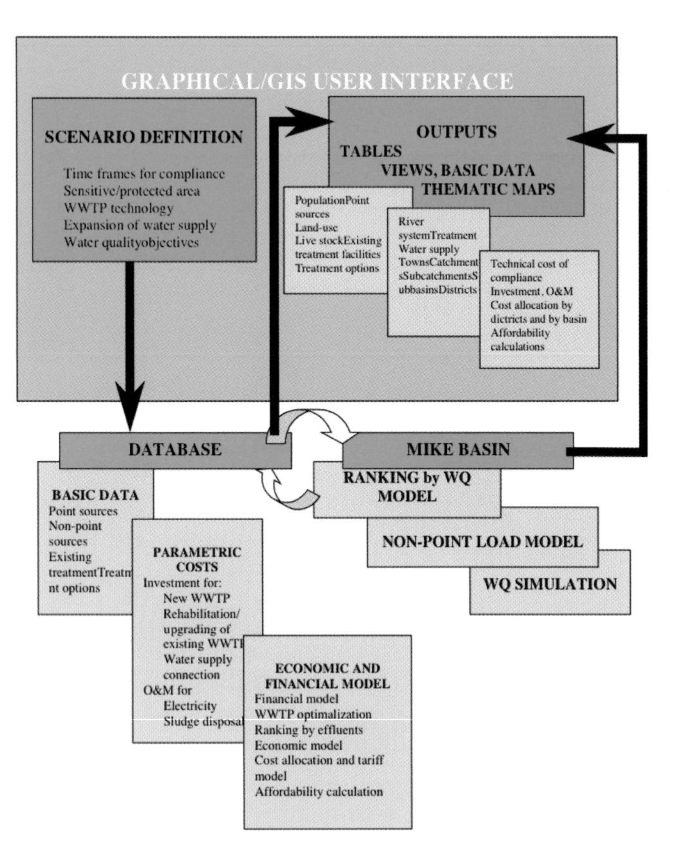

Components of the Decision Support System, which comprises databases and models, and which functions as an integrated and user-friendly tool, allowing for evaluation of alternative options for compliance with standards. It takes into consideration diverse elements such as legislative mandates, technical options, environmental impacts and economic implications.

Introducing MIKE BASIN

A key component of the DSS was the Danish Hydrological Institute (DHI) Water and Environment's MIKE BASIN, a software tool that combines the GIS functionality of ArcView with a hydrologic modeling capability. As the name suggests, this package operates at the river basin level, making it ideally suited for work associated with the WFD. MIKE BASIN was designed around a simple model for ease of use and to minimise the data requirements. Its sophistication is mainly in its visualisation tools, which permit the simulation results to be viewed both spatially and temporally. According to the DHI, this makes it ideal for building understanding and consensus.

MIKE BASIN models a river basin as a network in which branches represent individual river sections and the nodes represent confluences, diversions, reservoirs and water users. ArcView is used for editing the model and the interface was expanded to allow network elements to be edited simply by right-clicking.

Of particular importance to the Czech Decision Support System was MIKE BASIN's Water Quality (WQ) module, which simulates the transport and chemical reactions of the most important substances that affect water quality. Chemical reaction assessment includes an evaluation of the natural breakdown of individual substances as well as the chemical reactions that occur among them. Many of these reactions are defined as first-order differential equations for which the user can either supply the constants or use default values. Included in these evaluations are levels of ammonia, nitrate, oxygen, total phosphorus, *E. coli*, chemical oxygen demand (COD), and biological oxygen demand (BOD), as

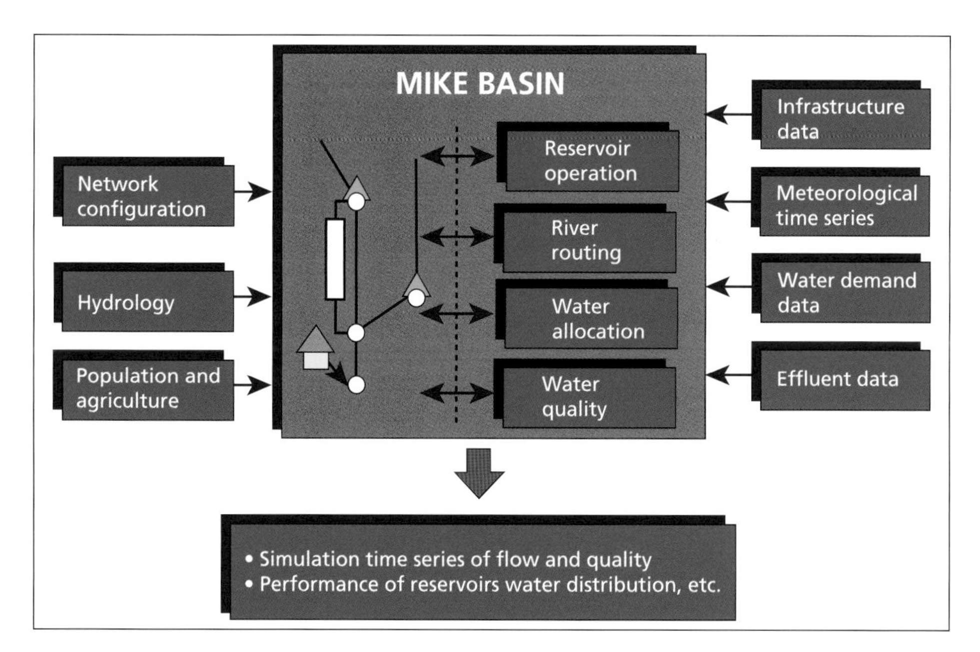

MIKE BASIN is a key component of the DSS and will allow the general user, such as regional planners and civil servants, to perform large-scale analyses without requiring extensive experience in databases, GIS, or modeling tools.

well as evaluation of user-defined criteria such as salinity. Additionally, the model takes into account temperature-contingent concentrations, re-aeration from weirs and time-dependent oxygen sources and sinks, such as respiration and photosynthesis. Sources of pollution—both point and nonpoint—and water treatment methods are modeled. The user can specify a method of water treatment or employ a default method.

Varied scenarios

The project to meet the European Water Framework required evaluating how the DSS would model three scenarios, since the exact criteria had not been clarified at the time. Two scenarios applied to the country as a whole, while the third included only sensitive areas. The latter were defined as all river basins where drinking water supplies and recreation-designated reservoirs were located.

The catchment areas of all river reaches were used as a source for water supply exceeding 500 000 cubic metres per year. According to this scenario, the sensitive area would encompass 61 percent of the whole country.

For each scenario, various strategies were modeled. Some strategies involved building new water treatment plants, while others meant upgrading existing plants. Strategies also differed in the method of sludge disposal (landfill or incineration), on the question of whether each town should have its own water sanitation plant or should share a regional facility, and on whether Czech or EU standards were to be adopted for new sewer connections.

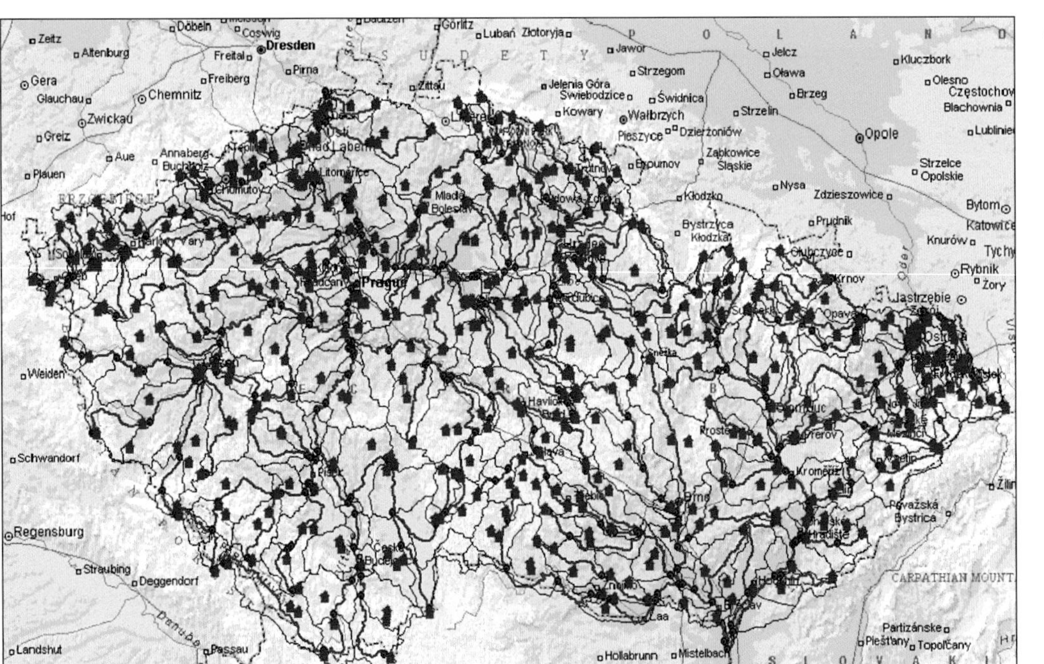

All major rivers, water usages and point and nonpoint source discharges in the Czech Republic are included in the water quality analysis of the DSS.

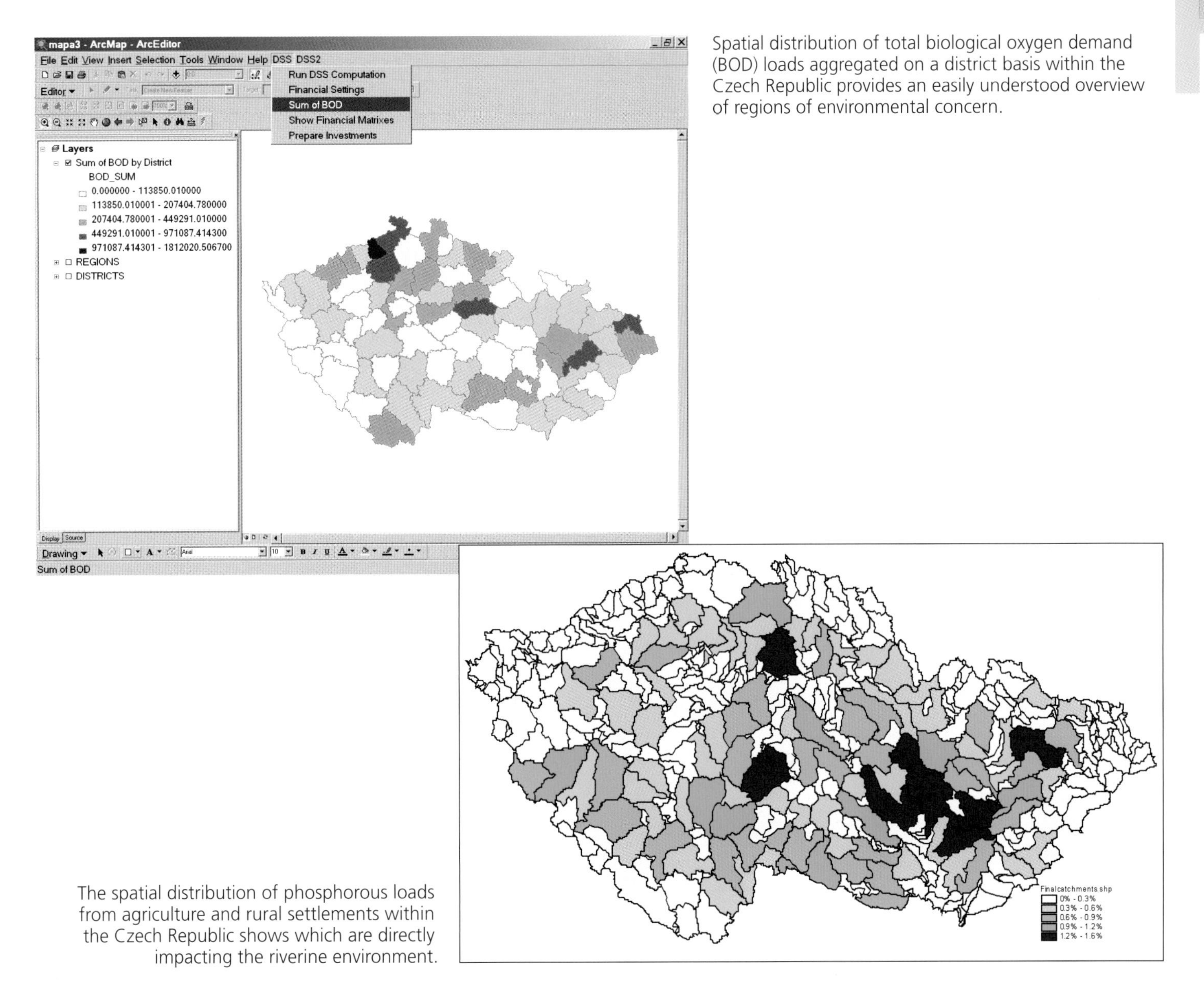

Spatial distribution of total biological oxygen demand (BOD) loads aggregated on a district basis within the Czech Republic provides an easily understood overview of regions of environmental concern.

The spatial distribution of phosphorous loads from agriculture and rural settlements within the Czech Republic shows which are directly impacting the riverine environment.

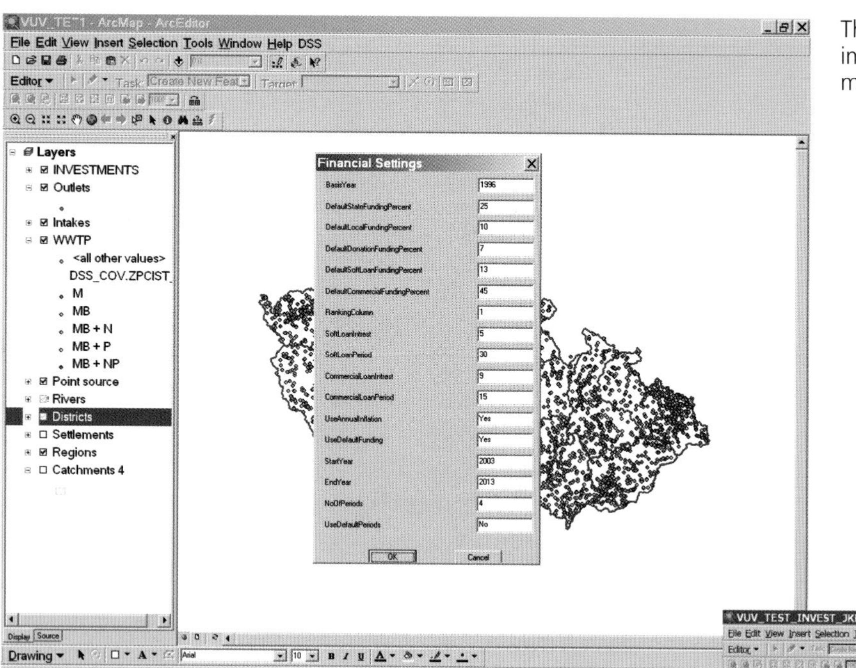

The financial investments of various environmental improvements are calculated by use of the economic models of the DSS.

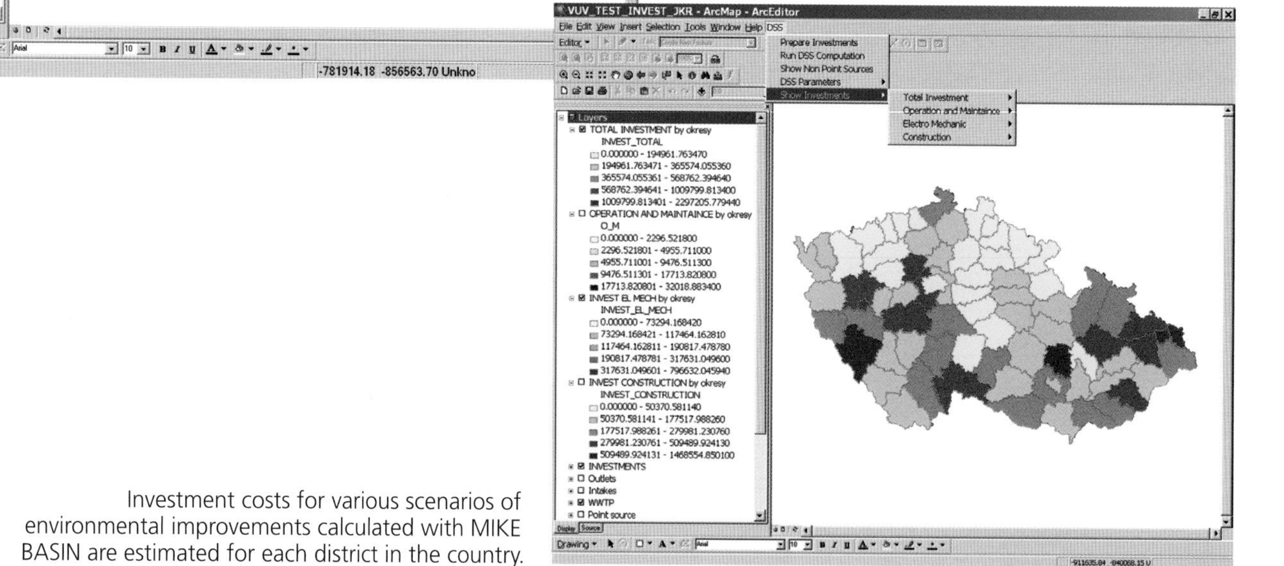

Investment costs for various scenarios of environmental improvements calculated with MIKE BASIN are estimated for each district in the country.

Presenting the results

The lowest-cost strategies for meeting an intermediate compliance target were determined for each of the three scenarios. The investment was not uniform across the country—costs differed from one district to another depending on the existing amount of pollution and the effectiveness of existing water treatment facilities.

To present the environmental benefits of each intermediate investment program, the change of water quality in any selected river branch over time can be displayed in what is known as a reduction curve, which can display an assessment of any water quality indicator in any part of the entire river system. When displayed with financial performance, the relationship between environmental benefits and total investment can be shown.

In order to present environmental benefits of all scenarios within the Czech Republic, the maps of water quality for the entire river network were generated directly from the DSS. All river branches were included and the water quality as it relates to a selected pollutant was indicated using different colours.

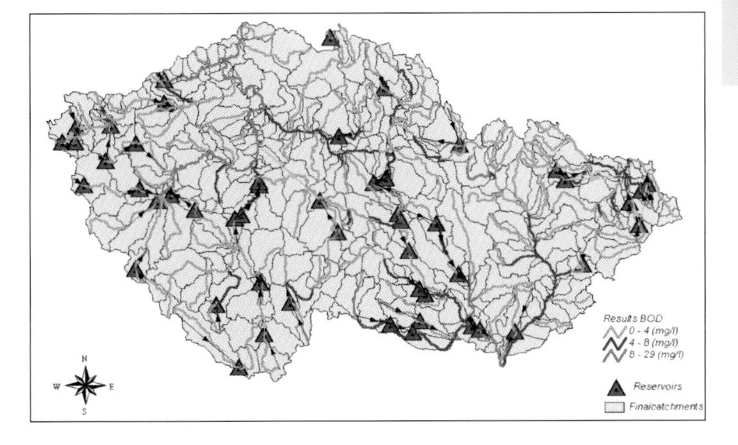

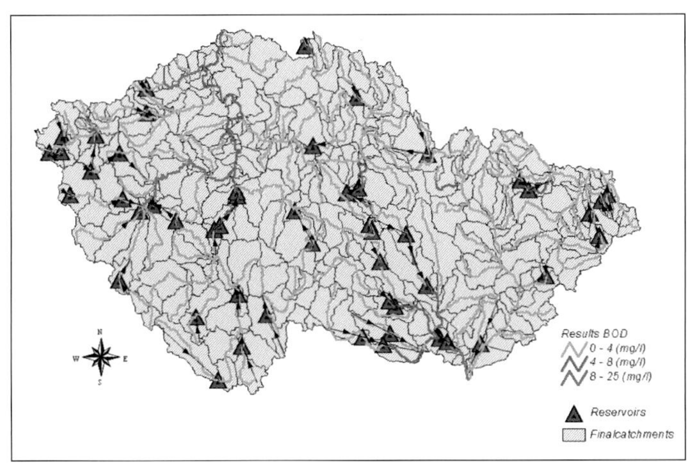

BOD water quality maps of Czech rivers for the third scenario of the DSS. Top: the baseline status. Bottom: the quality attained after implementation of a specific set of investments.

Hardware

PCs with Microsoft Windows® 98, Windows 2000, Windows NT, Windows XP operating systems. As a general guideline, a minimum of 128 MB DRAM and 1 GB of free disk space is recommended for MIKE BASIN. The simulation times depend on the CPU; a 200-MHz Pentium® II is the minimum recommended.

Software

The DSS was developed as a C++ extension on ArcGIS 8.3 and is based on a set of COM objects for the economic and financial computations. The MIKE BASIN computational engine is called from DSS once data is prepared. The results are presented back in DSS using ArcGIS 8.3 routines.

Data

Data sets include pollution sources (municipal, industry and nonpoint), receiving waters, existing water quality and hydrological conditions, water supply and wastewater treatment facilities, technical options for improvements, basic statistical data and topographical data, stored in a Microsoft Access 2000 database.

Acknowledgments

Thanks to Jan Krejcik, Stanislav Vanacek and Tomas Metelka of DHI Hydroinform a.s., Prague.

4 Revitalizing economic water resources

Two hundred years ago in Great Britain a far-reaching economic change began, when the largely agricultural economy began shifting to one based on manufacturing—a shift that would become known as the Industrial Revolution.

Although in time it would impact the entire planet, Britons felt its effects initially. One of the first areas affected was transportation; a manufacturing economy must have a reliable transportation system to flourish. Great Britain's system of canals was born from this need, and eventually, 4250 miles of waterways would crisscross the English countryside, bringing raw materials to manufacturing sites and finished products to market.

Today, the United Kingdom's canal system is not quite as critical to the economy as it once was, but it nonetheless remains an important resource for recreation and other activities. After years of decline, during which many canals fell into disuse or were even filled in, the potential of this element of Britain's industrial heritage is being reexamined. Many canals have been restored, and thousands of Britons are enjoying canal-related recreational activities such as boating, canoeing and fishing; the towing paths alongside the canals once trod by horses now serve as fine walking and bicycle paths.

In turn, this regeneration is providing an important boost to the tourism industry and helping rejuvenate many of the city centres through which the canals run, including Manchester and Birmingham. Economic activity is also reappearing: a fibre-optic network has been laid under 800 miles of canal towing path, and Watergrid—a public–private partnership of water agencies—is using sustainable canal water resources to supply waterside industries and property developments. All this activity shows that a two-hundred-year-old national resource needn't languish in the nineteenth century.

Although Britain's canals fell into disuse after the opening of the railways, today they are thriving once more thanks to tourism.

British Waterways, the public corporation that now manages the canal system, has recently implemented an Enterprise Resource Planning System (ERP) from SAP, which integrates all areas of its business. Since the resource being managed is geographically diverse, GIS is a key element of British Waterways' new system, used as a front end into SAP applications and linking many of those applications. This chapter examines two of British Waterways' GIS-enabled applications: day-to-day monitoring and system control, and hydrologic modeling.

Monitoring usage with hydrologic models

The British Waterways Water Resource Model was developed by Associated British Ports Marine and Environmental Research (ABPmer) and provides decision support for short-term operational planning of water resource use. It's also used for long-term strategic planning of water supply and transfers, abstraction licence agreements (water taken from canals and used by canal-side business), statutory water releases (water released back into rivers), and assessing the impact of climate change and new infrastructure,

Traditional narrow boats, the English countryside, canal-side pubs and industrial museums are drawing visitors by the thousands to the waterways.

such as marinas. In order to understand the modeling, we need to understand some of the elements of a canal system.

The use of locks to raise or lower boats from one section of a canal to another is well known. Less well known, perhaps, is the fact that using a lock transfers a considerable volume of water down the canal. Lock operation is only one of many actions, including the use of abstractors and statutory water releases, that can have this kind of serious effect on the total volume of water in use in a modern canal system. These high-volume demands are met from supply sources such as reservoirs, feeders, groundwater sources and back pumps that pump water back up the canal system.

The resurgence of interest in the canal system is also paving the way for the renovation of many inner-city areas through which the canals run.

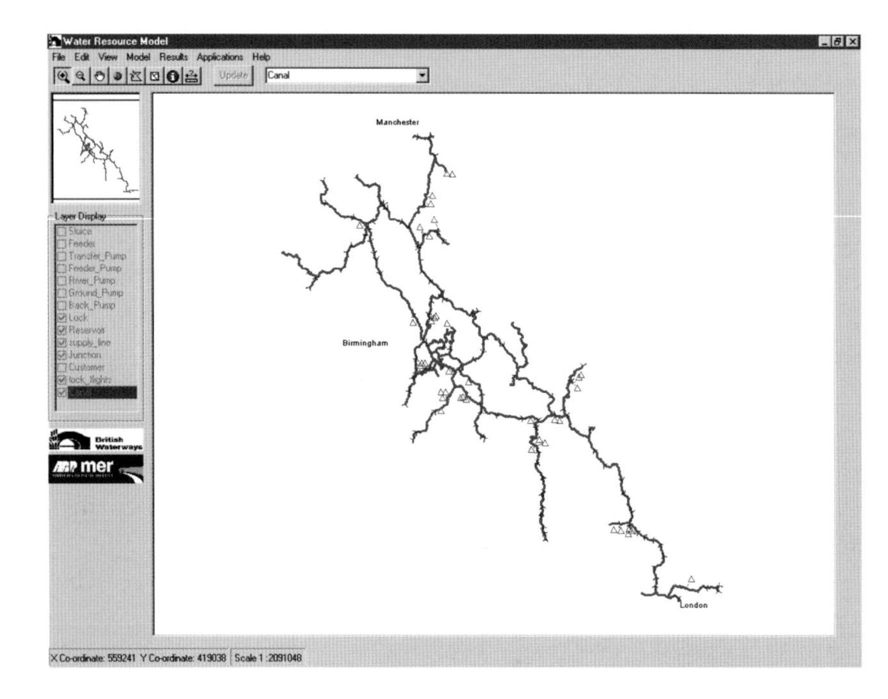

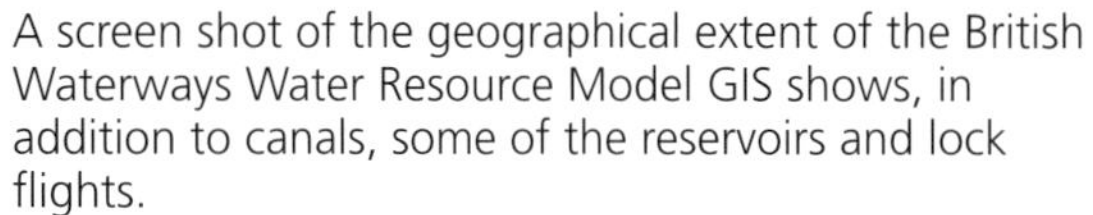

A screen shot of the geographical extent of the British Waterways Water Resource Model GIS shows, in addition to canals, some of the reservoirs and lock flights.

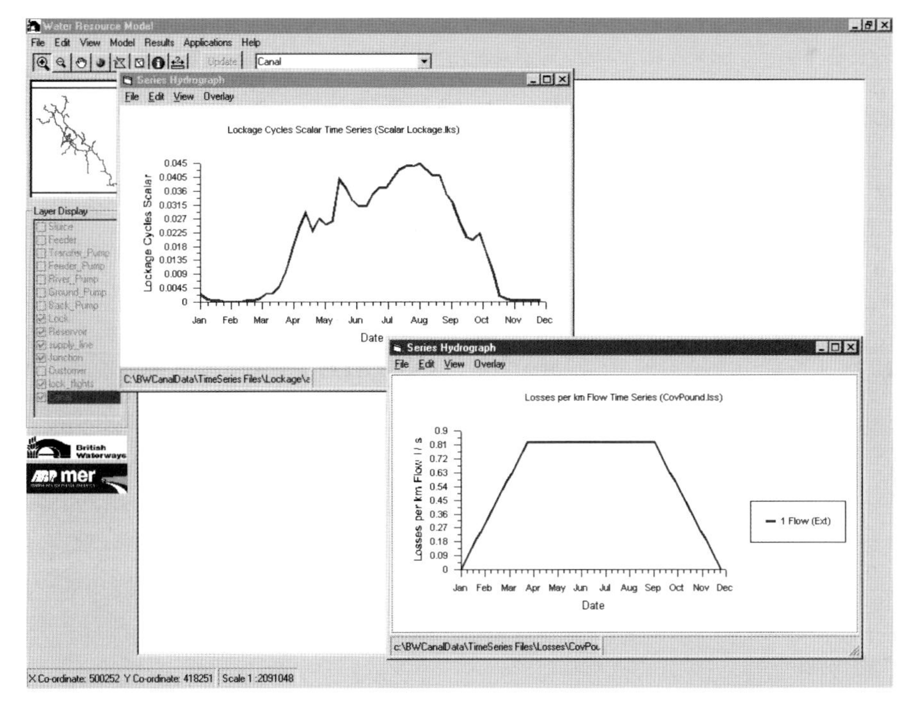

The annual distribution of demands from lockage and losses. Lock usage is highest in the summer due to tourism. Losses also peak during the summer due to climatic and operational factors.

The hydrologic model balances demands against supplies, and the hydraulic capacity of the canal system to move water from source to demand zone.

Monitoring costs

Although reservoirs are the main supply sources in the system, back pumps, feeder inflows or river abstractions, and groundwater pumping also augment water supplies. All these methods have costs associated with their use, and the value of the source water itself is another economic consideration. Pumping is usually the most expensive option, so where possible, feeders and the reservoirs are used before pumps are. However, the reservoirs must not be overused to the extent that they would be unable to refill sufficiently in subsequent years. To minimise pumping costs while preserving reservoir levels, a level-dependent planning cost is assigned to water from reservoirs; that is, the lower the level in the reservoir, the higher the cost. A fixed planning cost is assigned to the pumps and all other sources. The so-called reservoir cost curves are used to compare the cost of pumping to the cost of using the reservoir in any given period, so that the most cost-effective solution can be chosen.

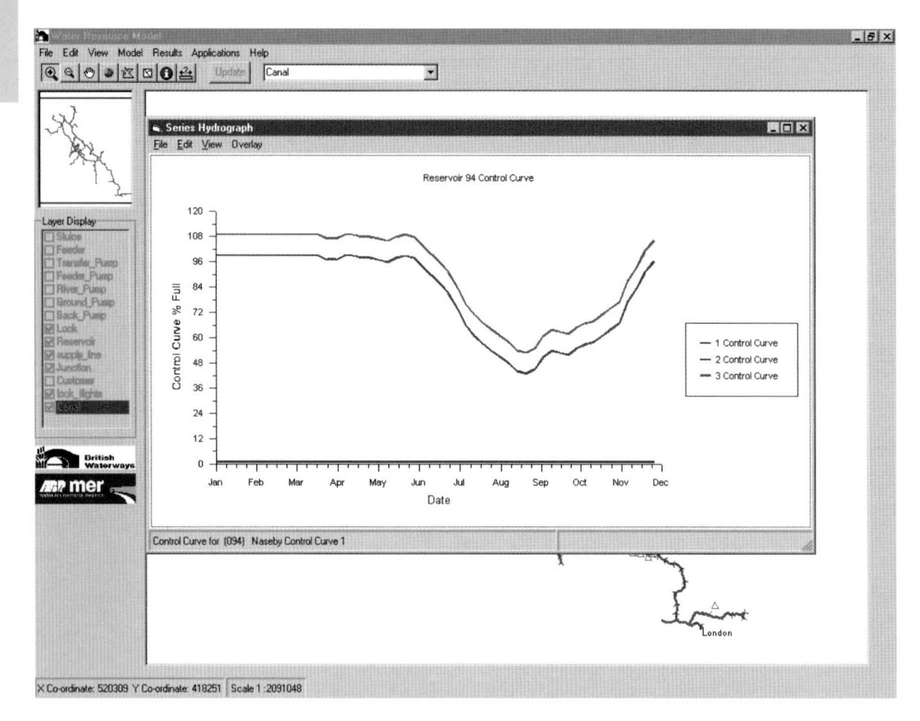

As reservoir levels decrease, the cost of water increases along these reservoir cost control curves. Reservoir water is valued relative to other sources to ensure optimal deployment of resources.

The optimal operation of the system depends on the correct allocation of these costs and particularly the cost of the reservoir relative to the cost of the pumps. A hydrological model allows these costs to be finely tuned. Other data is also available to refine the calculations, in this case, 80 years of historical reservoir inflow and feeder flow data. This enables an assessment of system performance over a historical period, and particularly over periods of drought. It also enables testing under different climate scenarios.

Planning for climate change

Although British Waterways has historical rainfall and modeled catchment flow data going back to 1918, many scientists believe that past trends aren't necessarily representative of what we can expect in the future. One of the predicted effects of climate change is a shift to drier summers and wetter winters. Obviously this will have a major effect on reservoir levels and, correspondingly, the need for auxiliary supplies.

To plan for the future, the hydrological model will be used to assess the impact of climate change on the future resources of the system. As an initial assessment, the model has been used with rainfall data typical of what can be expected as a consequence of climate change, and it has been found that reservoir levels will, typically, be lower than their current levels in summer, while winter refill will be faster. In time the model will be used to investigate remedial measures such as the development of complementary sources to increase system reliability in the summer months, and methods to capitalise on the faster winter refill.

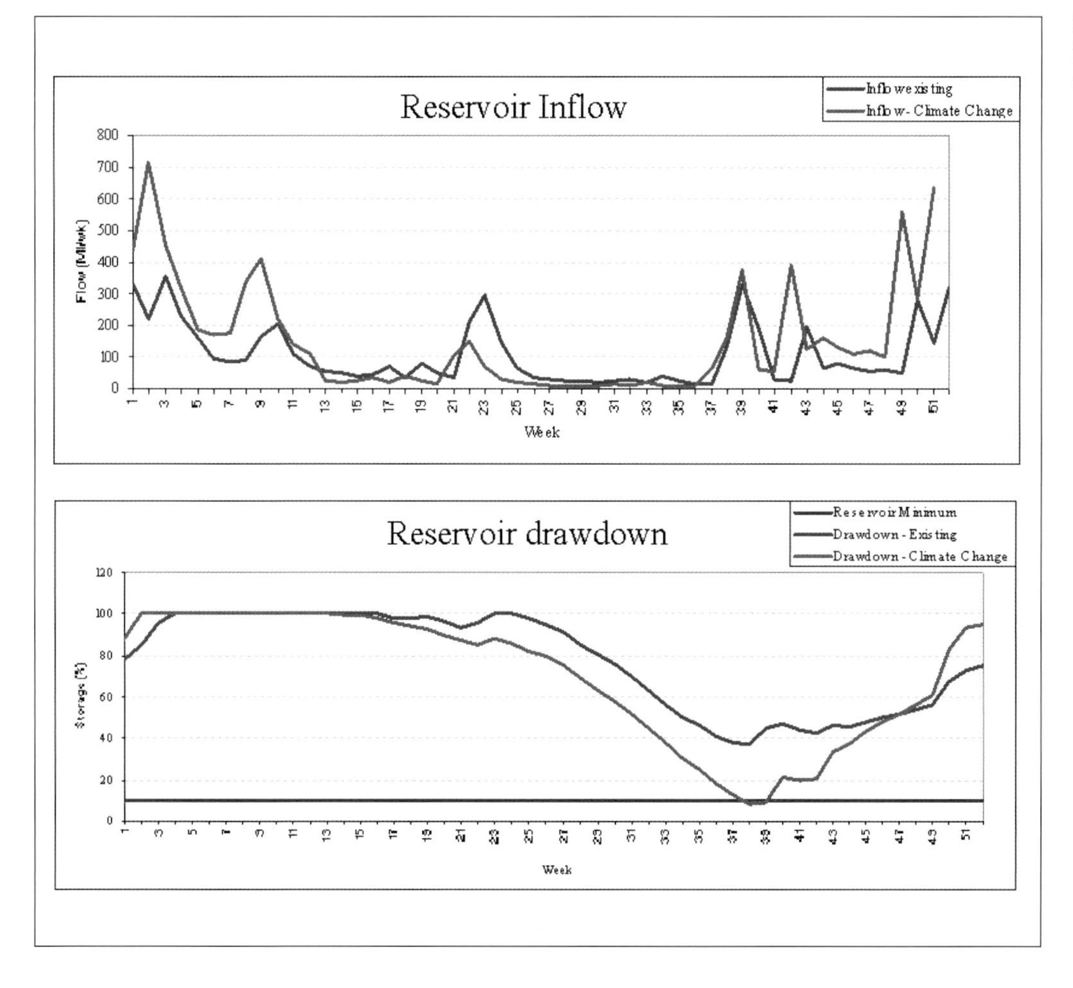

Reservoir inflow and drawdown for existing conditions and climate-change scenarios

A growth industry

Leisure activities in general, and the recreational use of the waterways in particular, are growth areas. To allow for this growth and to support the British tourist industry, British Waterways is continually assessing plans to extend the canal system. A significant element of the expansion programme is the building of new marinas. The additional traffic generated by new marinas will have an impact on the hydrology of the canal system; modeling provides a means of assessing this impact.

One British Waterways study examined the impact of the development of seven new marinas in one region of the canal network. Based on current usage patterns, the number of trips the extra boats would take is forecast and the increased lockage demand determined. By comparing the effect of individual marinas and of all proposed marinas against the current status quo, planners will be in a position to make informed decisions on new developments.

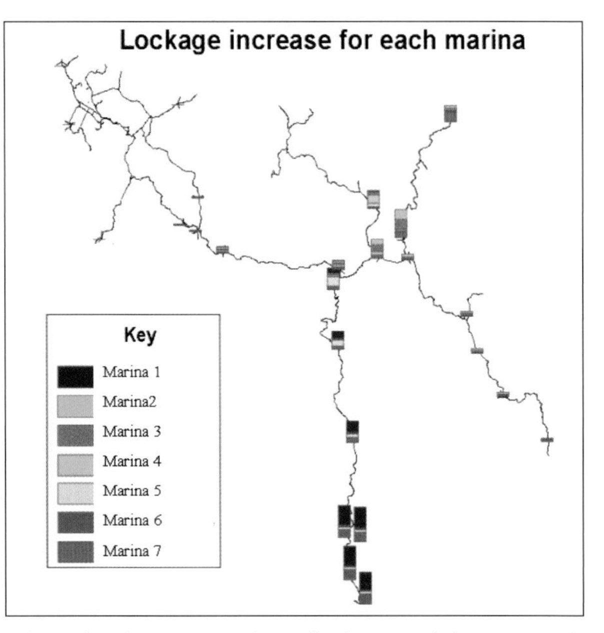

When planning new marinas, the impact of the proposed developments on water resources is assessed to ensure that the long-term reliability of the system is preserved. Here the spatial distribution of lockage increase for each new marina is shown.

The SCADA system

Monitoring the more day-to-day, real-time operation is achieved using another system accessible from the GIS/SAP combination—the supervisory control and data acquisition (SCADA) system.

Critical parts of the network are monitored using telemetry. Sensors in the canal and on equipment provide remote measurement of a variety of information including canal and reservoir levels, feeder flows, pump status and sluice gate positions. The SCADA/GIS interface enables the user to select an overview to see the status of a section of the network, including water levels, or to zoom in to view the status of an individual piece of equipment such as a pump.

In control

SCADA is used for more than just keeping an eye on the canal network—it also allows automatic control, mainly of pumps and sluices. By monitoring the necessary parameters, the system is able to start and stop pumps and operate sluices using the same operational knowledge that, in former times, a human operator would have employed.

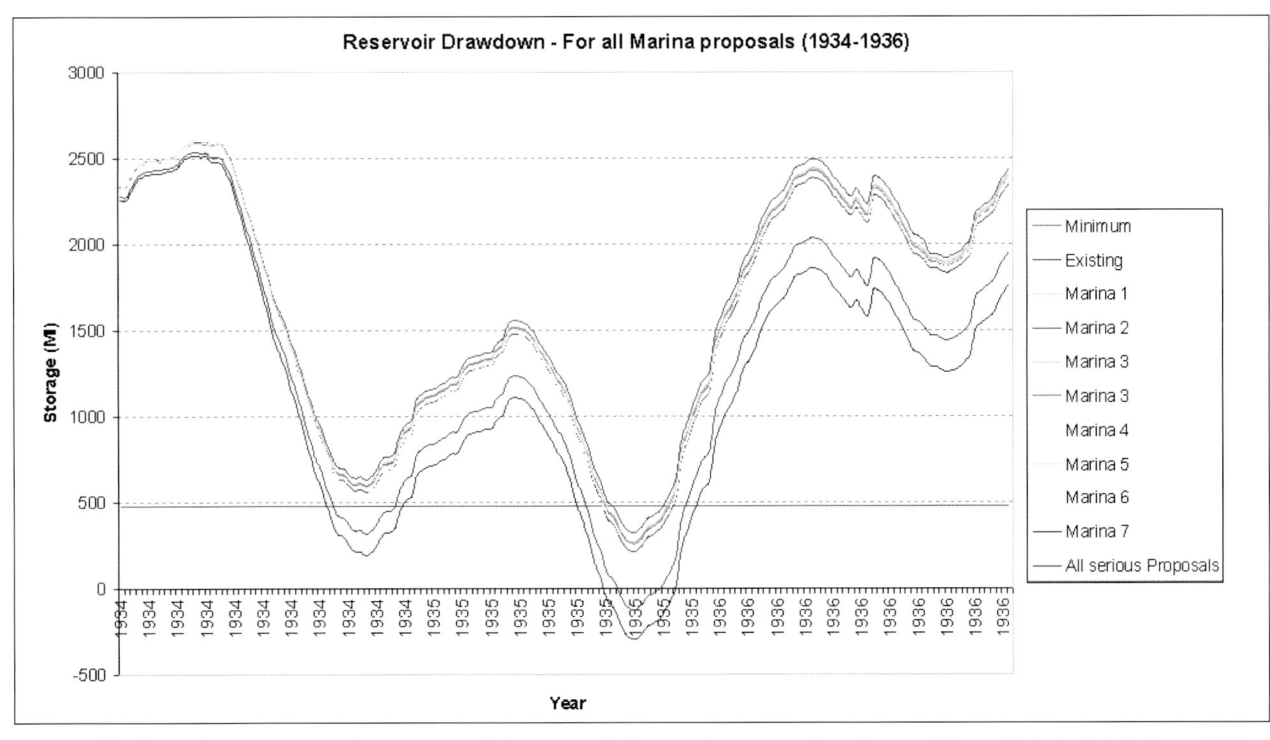

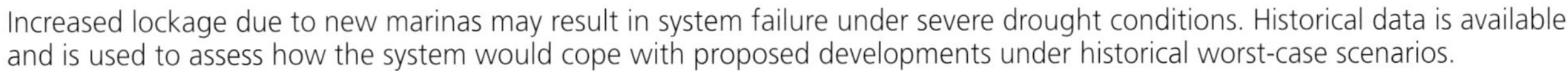
Increased lockage due to new marinas may result in system failure under severe drought conditions. Historical data is available and is used to assess how the system would cope with proposed developments under historical worst-case scenarios.

The automated approach, however, is more efficient, in terms of water usage, staff time and energy consumption. Whereas manually operated pumps, especially those in remote areas, are often left running unnecessarily, an automated pump is turned on only when it's needed. This can reduce energy costs by up to 50 percent.

Typically, the control scheme for a back pump will compare the current canal level with a target level, switching on the pump when the actual level is too low. Because energy costs are reduced at night, a lower level is allowed during the day. For a single pump this control regimen is comparatively simple, but extending this to a complete water transfer system comprising multiple pumping stations, and with the output of one feeding the input of another, requires considerable fine tuning.

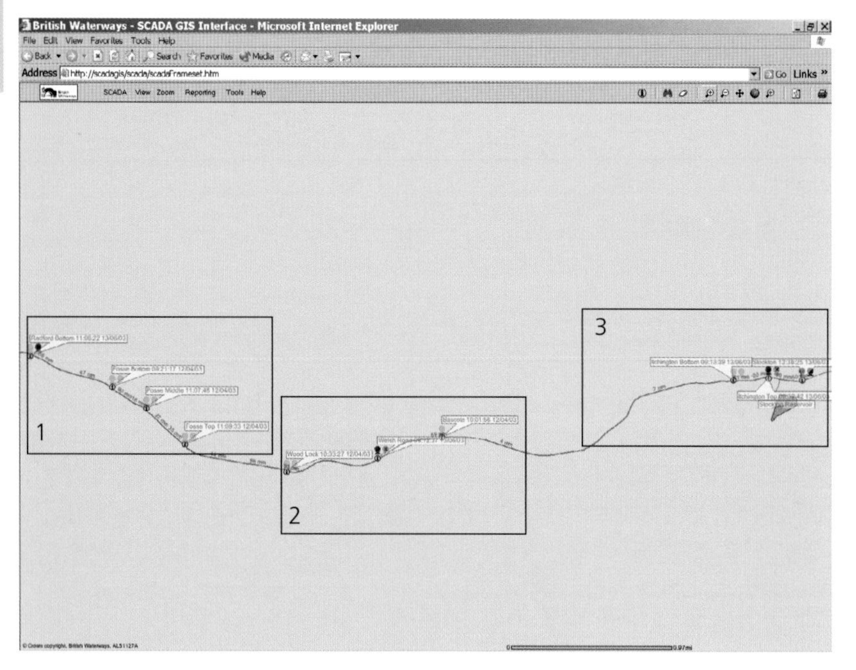

The SCADA system allows operators to monitor plants remotely. Here, 10 pumping stations have been linked to create a single water transfer route; the diagram shows that a pump at Welsh Road has failed.

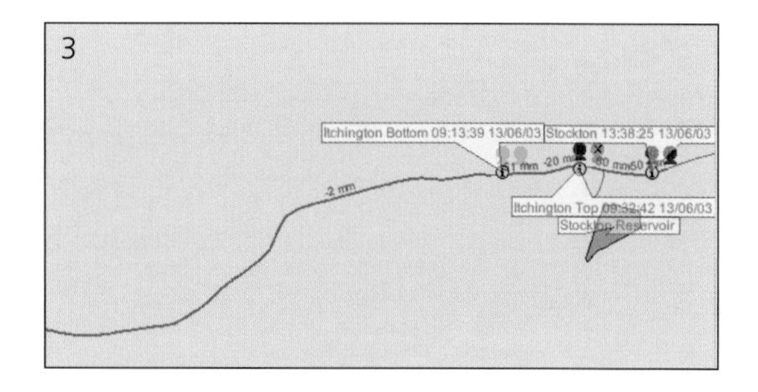

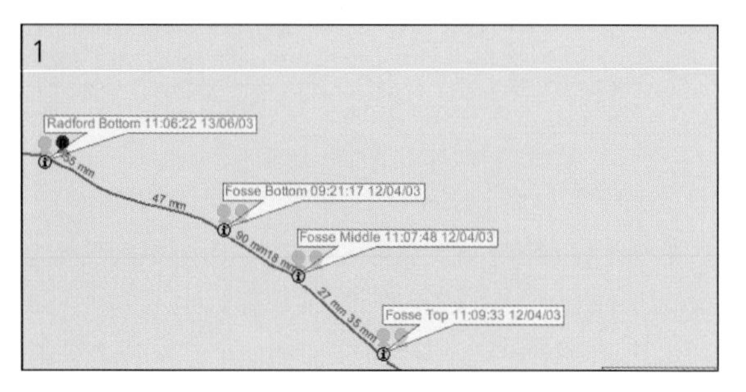

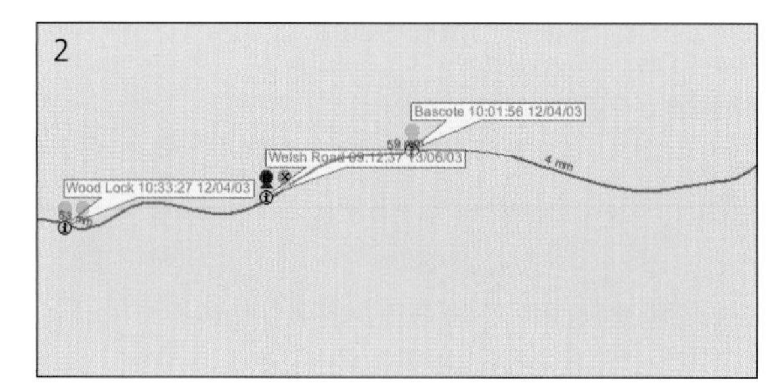

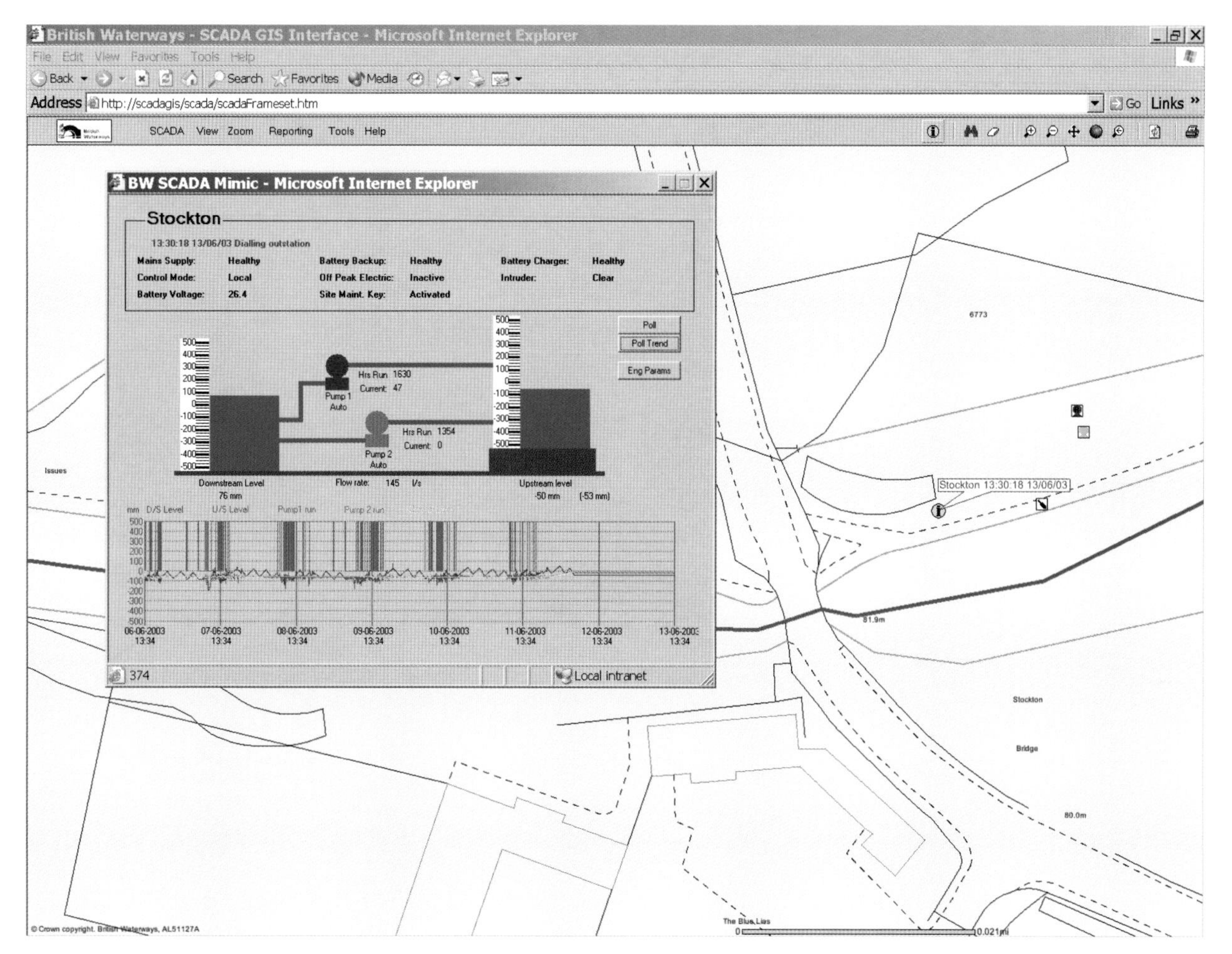

The Stockton pumping station is shown as a SCADA mimic. In this diagram of the plant, the status of the pumps is shown by colour-coding; live values, including water levels and pump currents, are indicated.

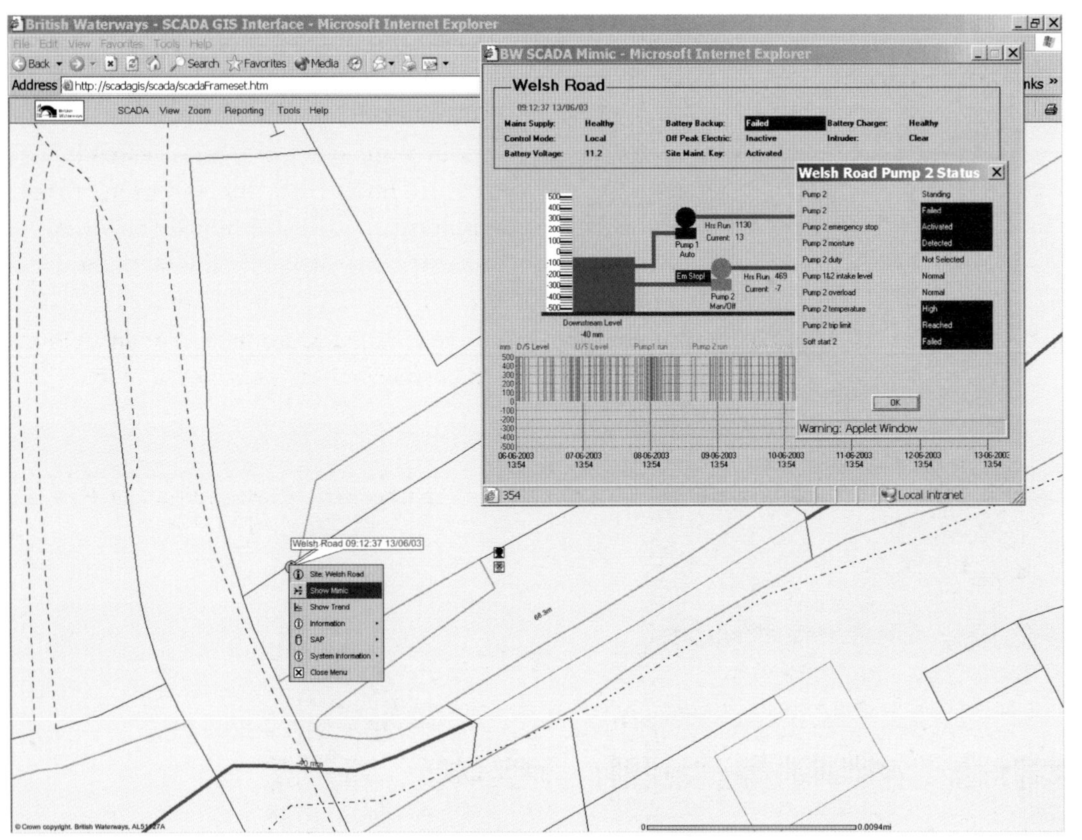

The SCADA mimic can be accessed from the GIS to find out more information on the Pump 2 failure.

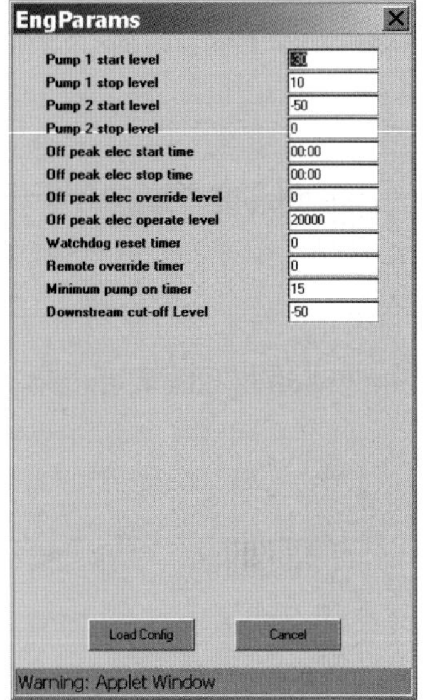

In addition to monitoring the system, SCADA also allows the plant to be controlled remotely. Here the parameters that dictate how the automatic control system works are being modified.

The information age

Up-to-the-minute information is essential to the efficient operation of the waterways. Before the introduction of the SCADA system, faults would become apparent only after the consequences of the problem manifested themselves. This would often be too late to avoid disruption or increased costs. With telemetry, however, equipment that has failed is immediately obvious. Telemetry also doesn't require human monitoring—the system provides warnings and updates via a whole range of interfaces, including pagers, mobile phones and PDAs.

Historical trends are also essential in reviewing the operation of the system and these applications can be launched directly from the GIS interface.

Although accurate and timely information is essential, it's only part of the story. Information serves a useful function only if action is taken on it. So an important element of the SCADA GIS interface is the integration with the company's SAP system. This allows an engineer accessing information on a piece of failed equipment to open an SAP session to schedule the relevant maintenance activity.

SCADA interfaces to British Waterways' SAP system so that maintenance can be scheduled for failed equipment. Here, a new pump is being installed following a failure.

Within a short period of time the new pump is in place at the bottom of the draft tube. Remote monitoring has allowed the pumping station to be up and running again with a minimum of delay.

Hardware

British Waterways used Compaq desktop PCs and Toshiba laptops for development, and Compaq servers were used for GIS and SCADA and database applications. Hosting of the SAP application was outsourced to LogicaCMG plc.

Software

Microsoft SQL Server™ 7 and 2000
ArcSDE 8.2
ArcIMS® 4
ArcGIS 8.2
MapExplorer 2, ESRI (UK) Ltd.

Data

British Waterways data sets include waterway centrelines, docks, reservoirs, ownership boundaries, property polygons, bridges, locks, aqueducts, tunnels, culverts, weirs, sluices, embankments, cuttings and SCADA sites.

British Waterways is part of the Central Government service level agreement, and currently holds the following Ordnance Survey data products: Landline, 1:10 000 raster, 1:50 000 raster, 1:250 000 raster, Meridian, CodePoint, AddressPoint, and Strategi.

Other data sets were provided courtesy of Environment Agency/Scottish Environment Protection Agency; English Nature/Countryside Council for Wales/Scottish Natural Heritage; British Geological Society; Department for Environment, Food and Rural Affairs (DEFRA); and local authorities.

Acknowledgments

Thanks to Margaretta Ayoung, Dave Boyle and Saul Davies of British Waterways.

5 Harnessing earth's power

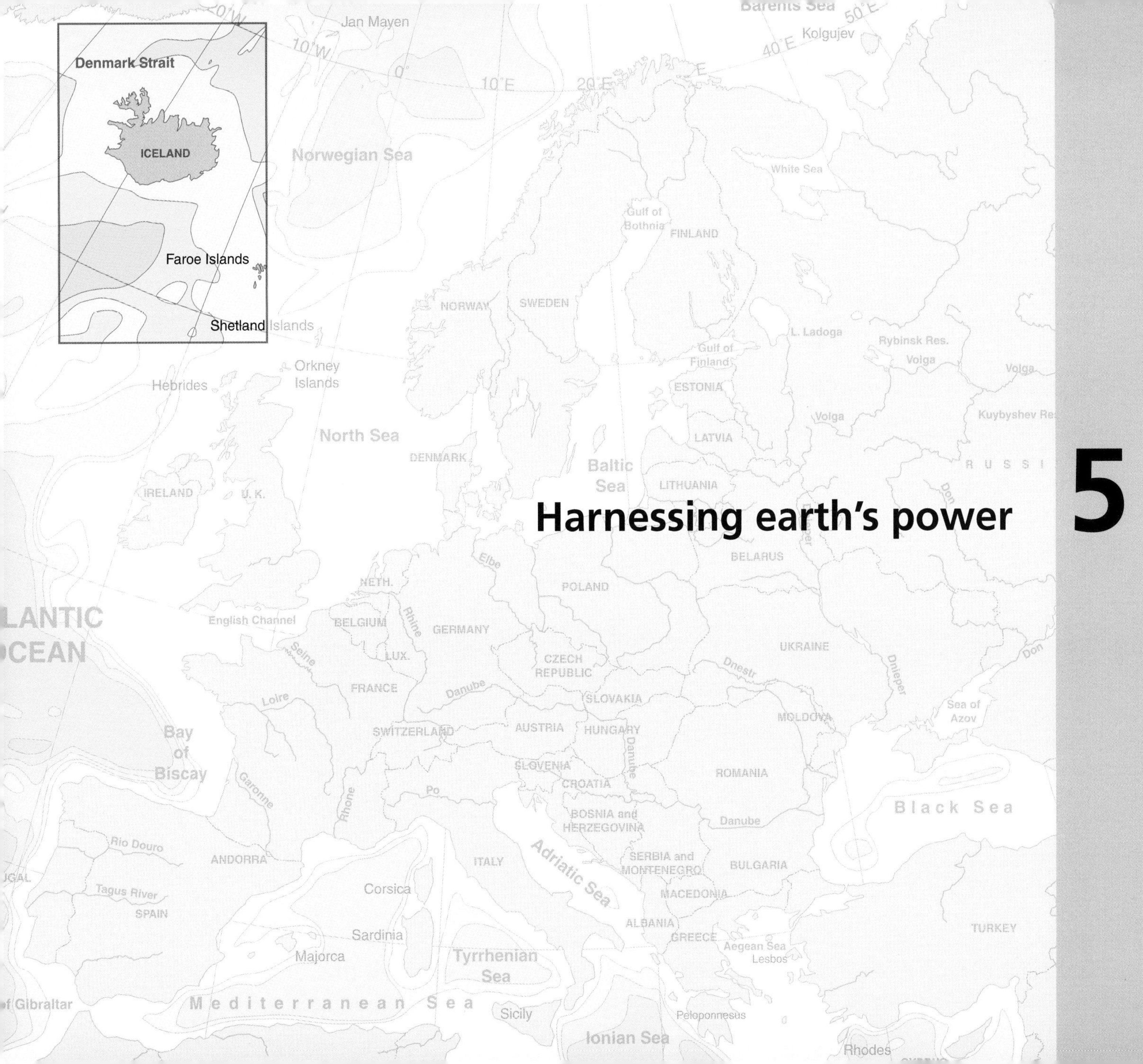

Translated literally, the name of the Icelandic capital Reykjavik means 'bay of smoke'—a phrase that might otherwise conjure up visions of a gritty coal town, or an industrial megalopolis. While it's true that Iceland consumes more energy per capita than any other nation, the reality is that very little of that energy is generated by burning fossil fuels. In fact, 87 percent of houses in Iceland are heated using geothermal water, vast reservoirs of which lie beneath the rocky Icelandic terrain. It is the mostly nontoxic vapors from these underground resources for which the city was named.

Today Reykjavik is one of the world's cleanest capital cities, thanks to geothermal energy.

Reykjavik is a city of contrasts: for example, even in July, the average temperature in Iceland is just 11° C. So homes must be heated throughout the year. For most of Icelandic history, its underground resources went untapped. Archaeological evidence suggests that the thirteenth-century saga writer Snorri Sturluson may have been the earliest user of geothermal hot water, but it's only been since the early 1970s that geothermal energy has overtaken fossil fuels as the main power source for home heating. Development since then has been rapid, however. With infrastructure in place, naturally occurring hot water is also being used in Reykjavik for generating electricity, filling swimming pools, melting snow from the roads, horticulture and fish farming, and miscellaneous industrial applications. Because some tasks, such as melting snow, can use water at a comparatively low temperature, the same water can sometimes be pressed into service for more than one job before it is discharged into the sea.

Reykjavik Energy provides geothermal water for the area from Borgarfjörður in West Iceland all the way to Hafnarfjörður, south of Reykjavik, and serves more than half the country's population. GIS helps manage Reykjavik Energy's heating system.

Prior to the early 1970s, Iceland's primary energy source was fossil fuels and Reykjavik suffered from smog just as many other industrial cities did.

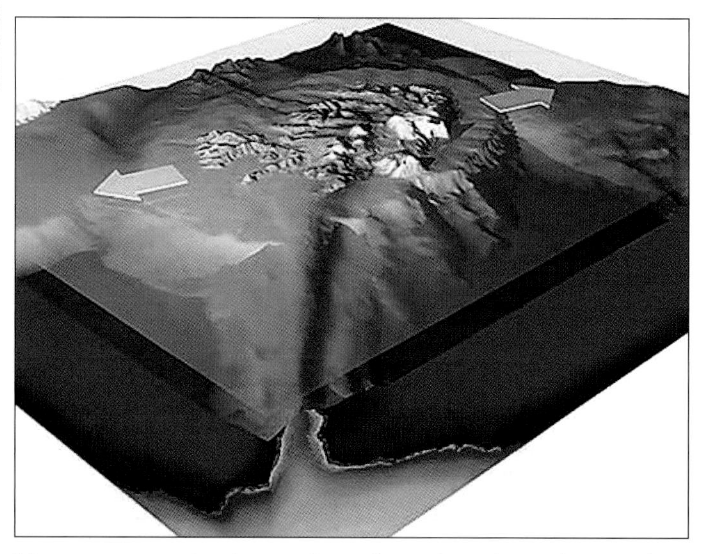

Magma comes quite close to the surface where the earth's crust has been disturbed by plate tectonics. The heat is transferred to water, creating a usable form of geothermal energy.

Eighty-nine percent of Iceland's 160 swimming pools are heated with geothermal energy. Most are outdoors and are in constant use throughout the year.

Geothermal heating: An overview

Reykjavik's geothermal water comes from three low-temperature geothermal fields and one high-temperature field. A low-temperature field is defined as one that contains water at below 150° C in its top kilometre. This water has few dissolved chemicals, so it can be used directly, either for heating or as tap water. Water from the high-temperature field—above 150° C—contains large amounts of gasses and minerals, so it can't be supplied to the consumer. Instead it's used indirectly, to heat cold water.

In Reykjavik's district heating system, low-temperature water is pumped out of the wells to deaerators that remove the small amount of dissolved nitrogen from the water. This prevents the buildup of gas in radiators that would stop the water from circulating. The water is next piped to storage tanks, then distributed to consumers. Storage tanks serve three purposes: they compensate for daily load variations, they provide peak energy during cold spells, and they provide hot water during power plant failures.

Even when it's been used for space heating and has cooled, the geothermal water still contains valuable energy. So it is stored in a separate set of tanks at 35° C. It is mixed with water coming from the wells to reduce that water's temperature to the 80° C required for space heating, and that water is stored in tanks at this temperature. Even the wastewater from this process can be used for melting snow.

Water from the high-temperature field at Nesjavellir is handled in a very different way. It heats cold water indirectly, through the use of a heat exchanger, which is then piped to the storage tanks. This heated freshwater is never mixed with the geothermal water from the low-temperature

The high-temperature field at Nesjavellir started producing hot water for the Reykjavík area in 1990. More recently, a power-generating plant has been added.

fields because its higher chemical concentrations would cause pipes to become blocked with substances such as magnesium silicate. This fresh, heated water is used to heat the southern part of the Reykjavik area.

Geothermal wells tap into Iceland's natural underground energy source.

These storage tanks feed the world's largest municipal geothermal heating service.

The distribution system encompasses 1300 kilometres of pipelines. Main pipelines like this one, which carry water from the wells to the storage tanks, are up to 900 millimetres in diameter.

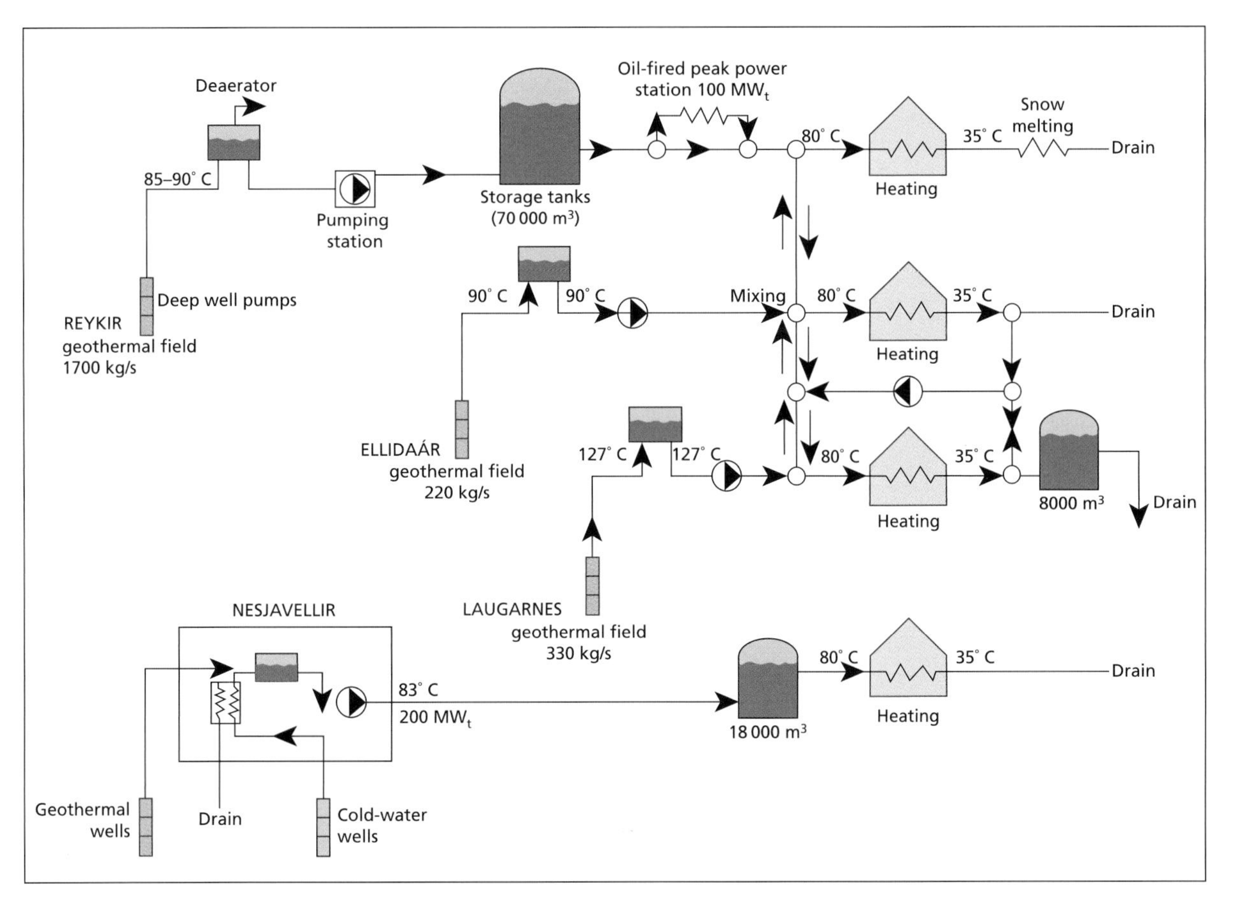

Reykjavik's heating system comprises a network of high- and low-temperature wells, storage tanks, valves and pumps.

Keeping track of pipework

Reykjavik Energy's GIS includes data for the district heating system, the electrical system and the cold-water distribution system. A major element of the heating system data relates to the distribution pipework, including information on the size of each pipeline, its role, its age and characteristics, and the network. The pipework makes up the majority of the installed plant, although the system also contains data on boreholes, wells, valves, pumps and storage tanks.

Throughout the development of the GIS, Reykjavik Energy has collaborated with the city of Reykjavik and

In this map, information on all underground utilities is shown, including information on the position and key characteristics of the various pipes and lines.

government-owned Iceland Telecommunications to form an enterprise GIS, overseen by an organization known as LUKR. Also used by the Public Works and City Planning offices, the enterprise GIS allows all these agencies access to information on streets, buildings and property boundaries, essential for work on the geothermal water distribution system. For their part, the Public Works and City Planning departments must keep track of their own 1300 kilometres of pipes that criss-cross Reykjavik.

Permission to dig

Digging in the streets of any city without detailed information about the underground services is a sure way to cause mayhem. But with the network of hot-water pipes in Reykjavik adding to the usual array of cold-water pipes and electrical, telephone and fibre-optic cables, the situation in Reykjavik is even more problematic. Needless to say, any construction work that involves digging in the streets has to be carefully monitored. GIS technology helps make this happen.

A schematic section of the Reykjavik hot-water system.

Contractors must obtain a permit from Reykjavik Energy to dig in Reykjavik's streets. The staff member handling the request finds the relevant site in ArcView, fills out a form with information about the details of the work to be done at the site, configures the maps and sends them to a printer or fax machine. The maps show all the relevant details of underground services in the dig location, even down to details about the connections inside a manhole. It is also possible to include the coordinates of any points that have been surveyed in the area in a fax or e-mail to the customer.

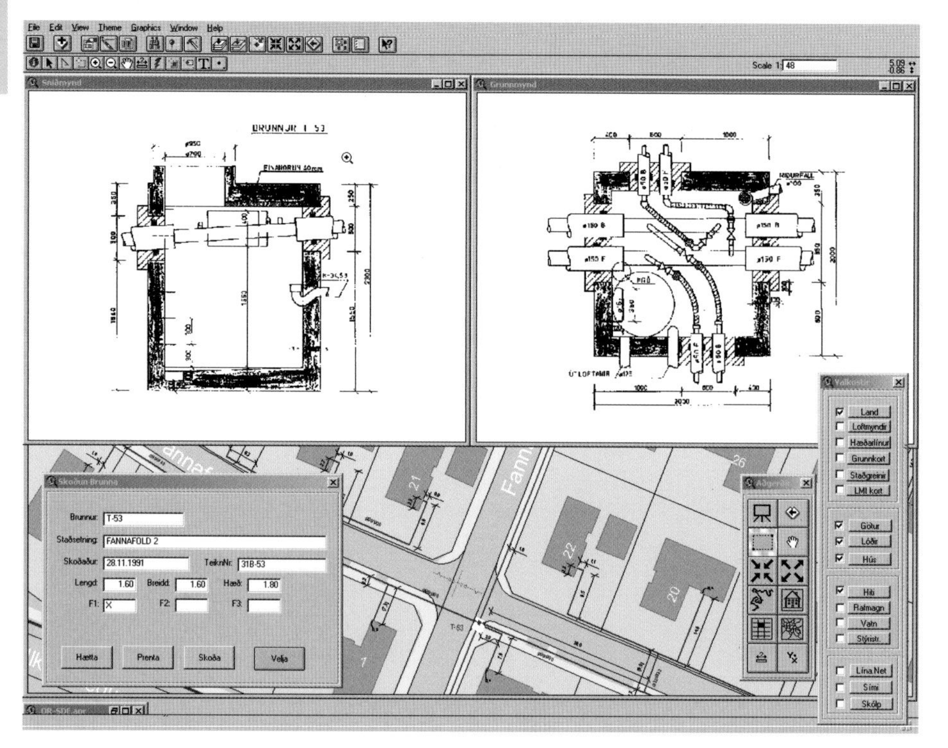

A legacy drawing of a well can be scanned in and viewed using GIS software.

Keeping up the pressure

To show something of the potential for GIS in Iceland, Reykjavik Energy has linked hot-water usage records for each house in the network to the enterprise system. By coupling this data on the diameter of the pipework and the elevation of each property, the hot-water pressure can be calculated at any point in the system. From this, a pressure contour map overlaid on a map of the pipework can be created. By identifying high-pressure regions, planners can see where additional users can be accommodated without adversely affecting existing customers.

In the field

Since faults can occur at any time, maintenance staff remain on call 24 hours a day. To ensure that they have the information necessary to carry out repairs after normal office hours, each of the construction vehicles is equipped with a laptop in the cab, and each can access a copy of current GIS data sets through ArcView installations on laptops they carry with them. This is useful in providing directions to the site and, once there, they can locate the relevant valves and so isolate the section of pipework they need to work on.

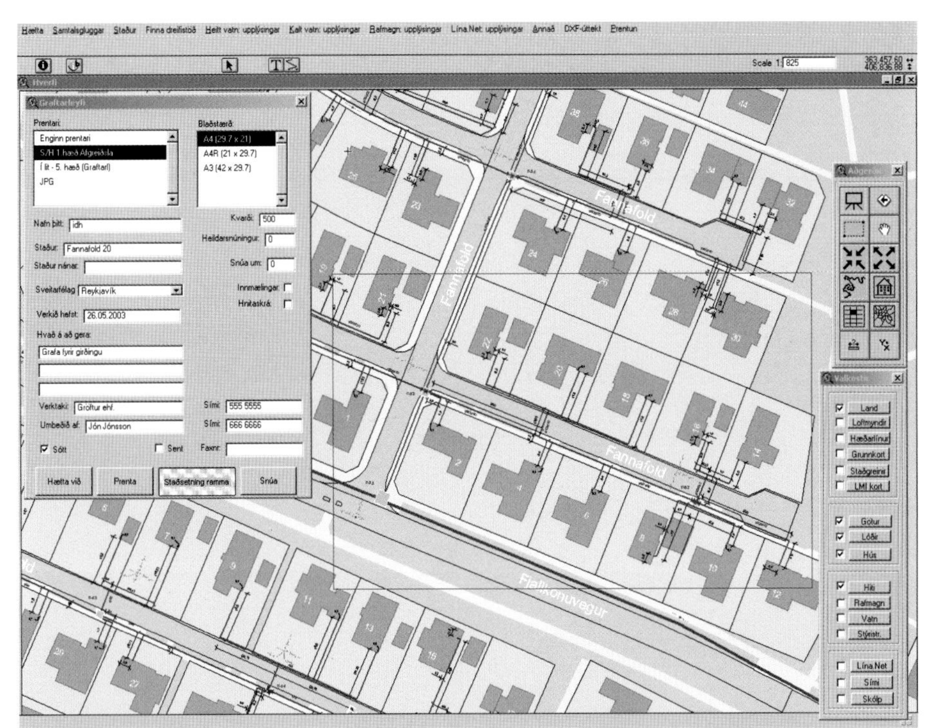

This dialog is used to produce *permission to dig* maps. The form is stored so that Reykjavik Energy can supervise the work if the need arises.

It's not just Reykjavik Energy employees who benefit from the ready availability of geographic information. The cooperation of various bodies in creating the GIS has resulted in some interesting spin-off applications. One such application is GIS use by the fire department to determine where hydrants are located, their capacity, and the capacity of waterlines. All of this information is available online in the fire trucks while the firefighters are driving to the incident. Using GIS data sets, the software can also calculate the total water supply available from the relevant hydrant and the attached hose.

Various types of analysis can be carried out on the data in the database. Here hot-water pressure is shown as contours; the relationship with pipe diameter is obvious.

Hardware

HP® dual-processor L3000 with 3 GB RAM, UNIX® OS.

Software

ArcInfo 8.2, ArcGIS 8.3, ArcView 3.2, ArcIMS 3, ArcSDE 8.2

Data

Reykjavik Energy's main databases cover electricity, the hot-water distribution system, the cold-water distribution system, fibre-optic cable, and land ownership of Reykjavik Energy. SDE® databases store information on the hot-water distribution system, including pipes, valves, survey points, pumping stations, storage, manholes, buildings and wells.

Acknowledgments

Thanks to Inga Dóra Hrólfsdóttir and Geir Svanbjörnsson of Reykjavik Energy.

6

Water is life at Pidpa

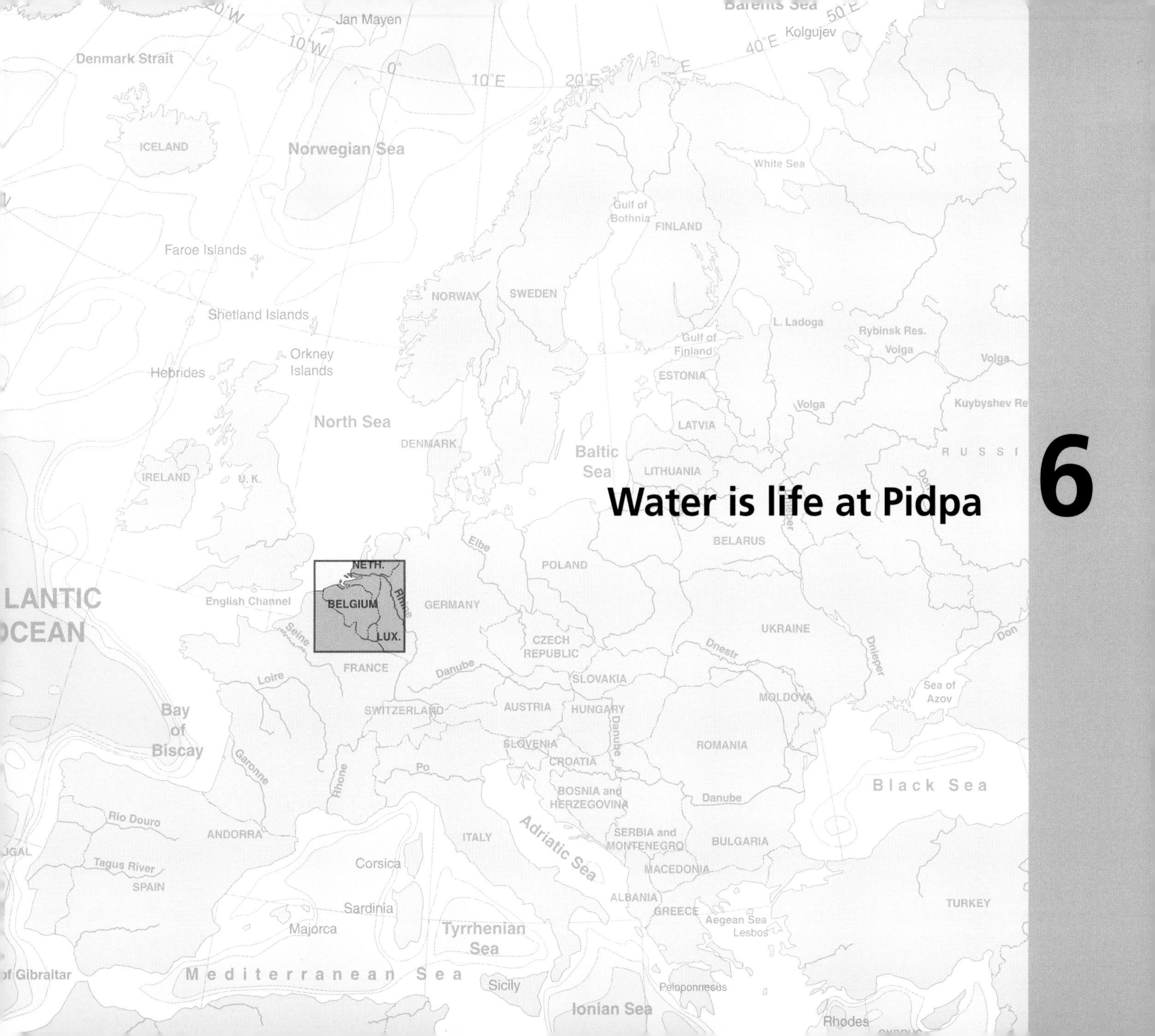

Water is life: a phrase so true that it has been adopted as the motto of one of the largest drinking water companies in the Flanders region of Belgium: Provinciale en Intercommunale Drinkwatermaatschappij der Provincie Antwerpen—Pidpa. The motto reflects the company's commitment to continual improvement in the quality of its water and in the service it provides its customers. Key to providing this service is a well-managed and well-maintained water distribution network. GIS is a critically important element of that goal. It is also key to a larger vision for the future.

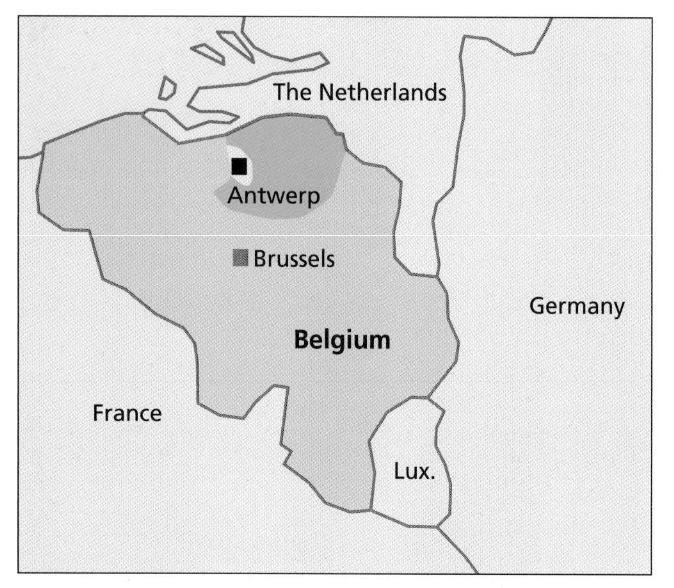

Centrally located in Europe, Pidpa provides high-quality drinking water to more than one million people throughout the Antwerp province of Belgium.

The efficient operation of any water distribution network relies on infrastructure, including water plants, water towers and pipework.

Pidpa's main office in the city of Antwerp houses more than two hundred employees working mainly in administration. About four hundred more are stationed throughout the province.

Each year Pidpa, a publicly owned utility, distributes more than 76 billion litres of water to more than one million people in 450 000 households and industries. Of this, 65,3 billion litres come from Pidpa's own 25 water production plants. The distribution network contains 62 water towers and more than 12 000 kilometres of water-main pipelines spread over an area of 2581 square kilometres.

Beyond the day-to-day work required to manage this far-flung enterprise, Pidpa has established links with other water providers in Europe and around the world. Their motive for making these connections is their strong belief in the usefulness of the GIS they have developed—they were the first water agency in Europe to adopt ArcGIS—and that their GIS tool will form an important element in a generic water utility tool for the entire European market. That tool, and a water data model specific to the European continent that could be shared among all European water organisations, are essential goals for the GIS program.

Pidpa uses several groundwater production facilities to ensure high-quality drinking water.

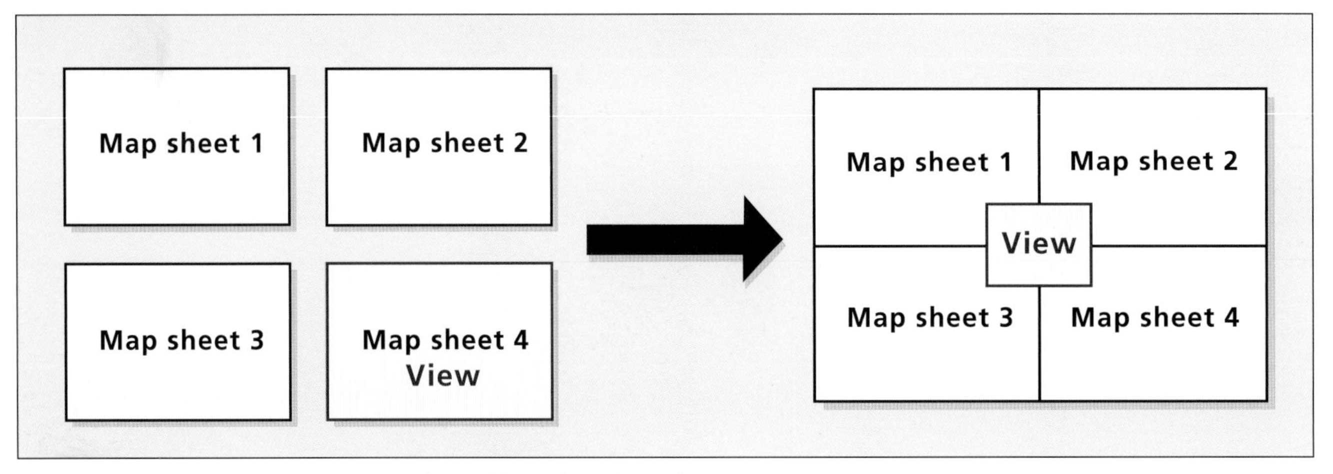

Using a seamless digital map overcomes the problems that arise with paper maps when the area of interest spans more than one sheet.

From paper to bytes

Before the creation of Pidpa's GIS, management of the Pidpa distribution network relied on a vast collection of paper maps and diagrams. Copies of about three thousand oversized A0-size maps at a scale of 1:1000, and 175 overview maps at 1:5000 scale, had to be stored in each of Pidpa's offices. There were also 90 000 A4-size synoptic drawings on hand that provided information about every element of the network. Keeping all these paper maps updated was a monumental maintenance problem, one that could be solved with GIS.

The first task was to convert all these reams of legacy paper data and maps into a suitable digital format. All the paper maps were scanned, processed and stored in a file structure that was georeferenced into a grid at scales of 1:1000 and 1:5000. This produced a single, seamless raster map onto which new network data could be added in vector form over several years. Data for a small portion of

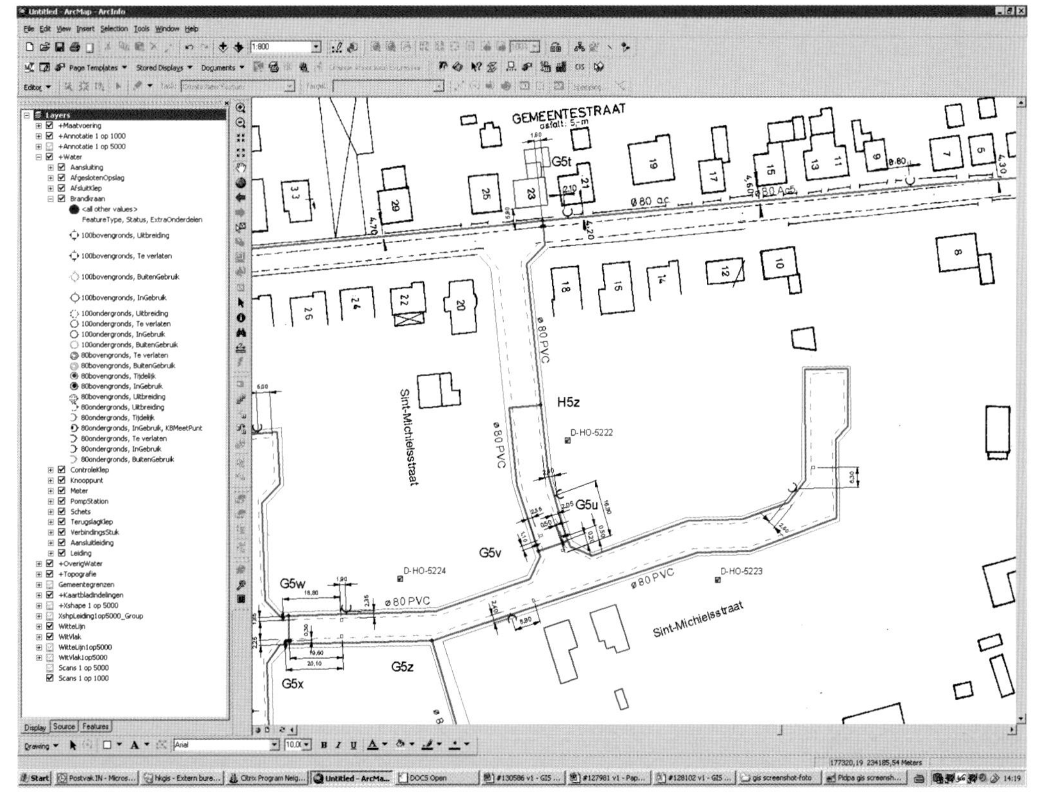

Historically, all of Pidpa's maps were on paper or film. Now these have all been scanned and seamlessly combined with vector data in an ArcGIS environment.

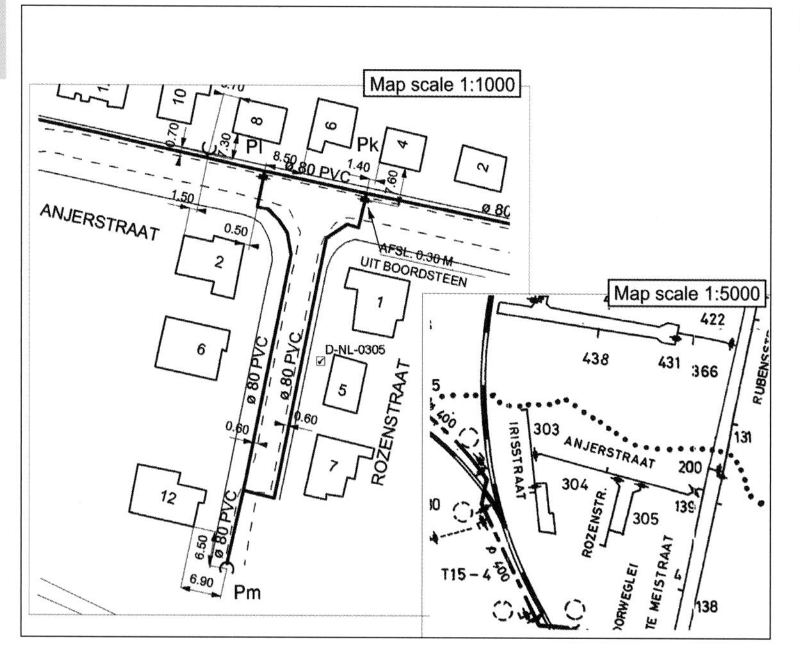

In the past, large numbers of maps at a scale of 1:1000 and 1:5000 were used. Here we see a scanned version of one of those maps showing the essential network data.

Pidpa's service area, around 7 percent, was already available in vector format from a previous GIS implementation; the conversion process imported this earlier data into the new system.

The paper maps at Pidpa contain topographic data as well as water network data. The lengthy process of vectorising the water network will be done by Pidpa or by a conversion company. The topographic data will be purchased from a Belgian governmental organisation as it becomes available over the next 10 years.

The 90 000 synoptic drawings—unscaled representations of network elements and connections—have been scanned to produce graphic files in TIFF format, which can be referenced by the GIS. But given the vector format's greater flexibility, Pidpa has launched the lengthy process of redrawing each of these diagrams in AutoCAD®.

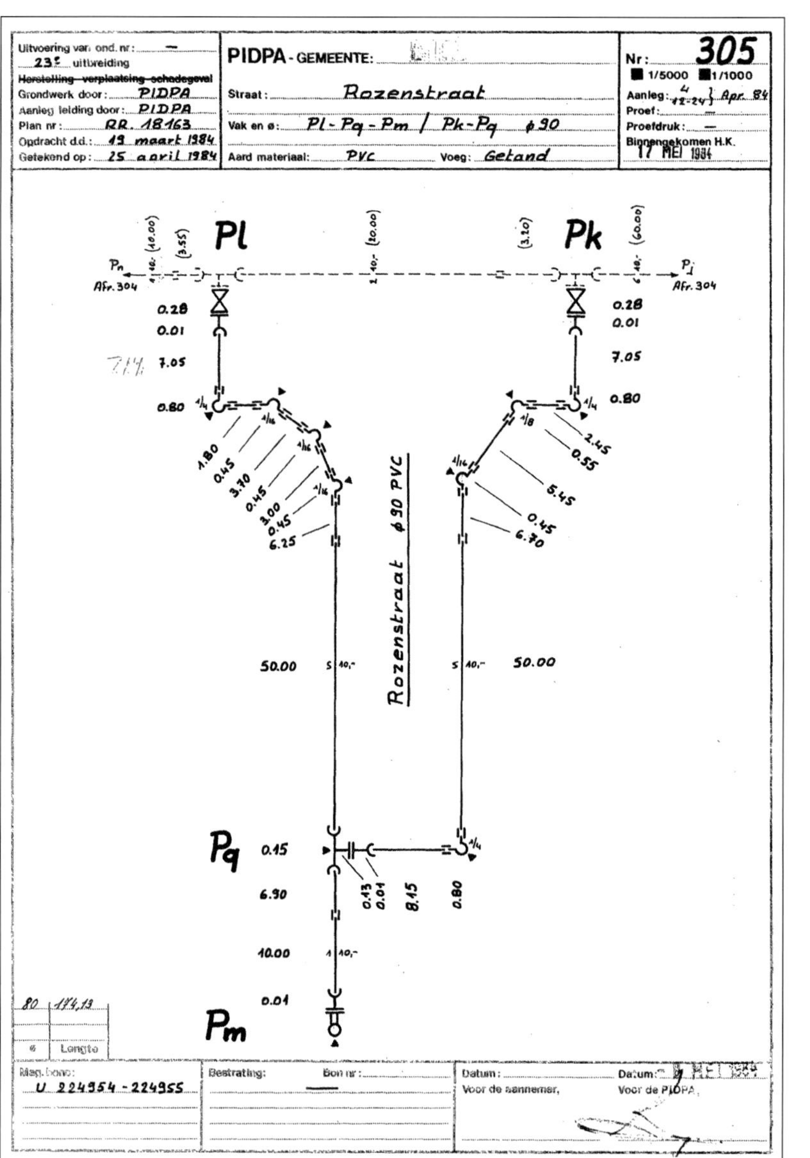

Uitvoering van ond. nr: —
23e uitbreiding
~~Herstelling - verplaatsing - schadegeval~~
Grondwerk door: PIDPA
Aanleg leiding door: PIDPA
Plan nr: RR. 18163
Opdracht d.d.: 19 maart 1984
Getekend op: 25 april 1984

PIDPA - GEMEENTE:
Straat: Rozenstraat
Vak en ø: Pl - Pq - Pm / Pk - Pq ø 90
Aard materiaal: PVC Voeg: Getand

Nr: 305
1/5000 1/1000
Aanleg: 4 / 12-24 } Apr. 84
Proef: —
Proefdruk: —
Binnengekomen H.K. 17 MEI 1984

Pl Pk Pq Pm
Afr. 304 Afr. 304
Rozenstraat ø 90 PVC
50.00 50.00

Ø Lengte
80 174,13

Map. bono: U 224954 - 224955
Bestrating: Bon nr:
Datum: Voor de aannemer,
Datum: Voor de PIDPA,

Detailed sketches provide additional information about the water network and are now linked to the maps at a scale of 1:1000.

To ensure fewer errors by users when they enter data, the Editor Extension to Pidpa's ArcInfo installation automatically checks to see if pipeline diameters match the parameters of the material used for the pipelines.

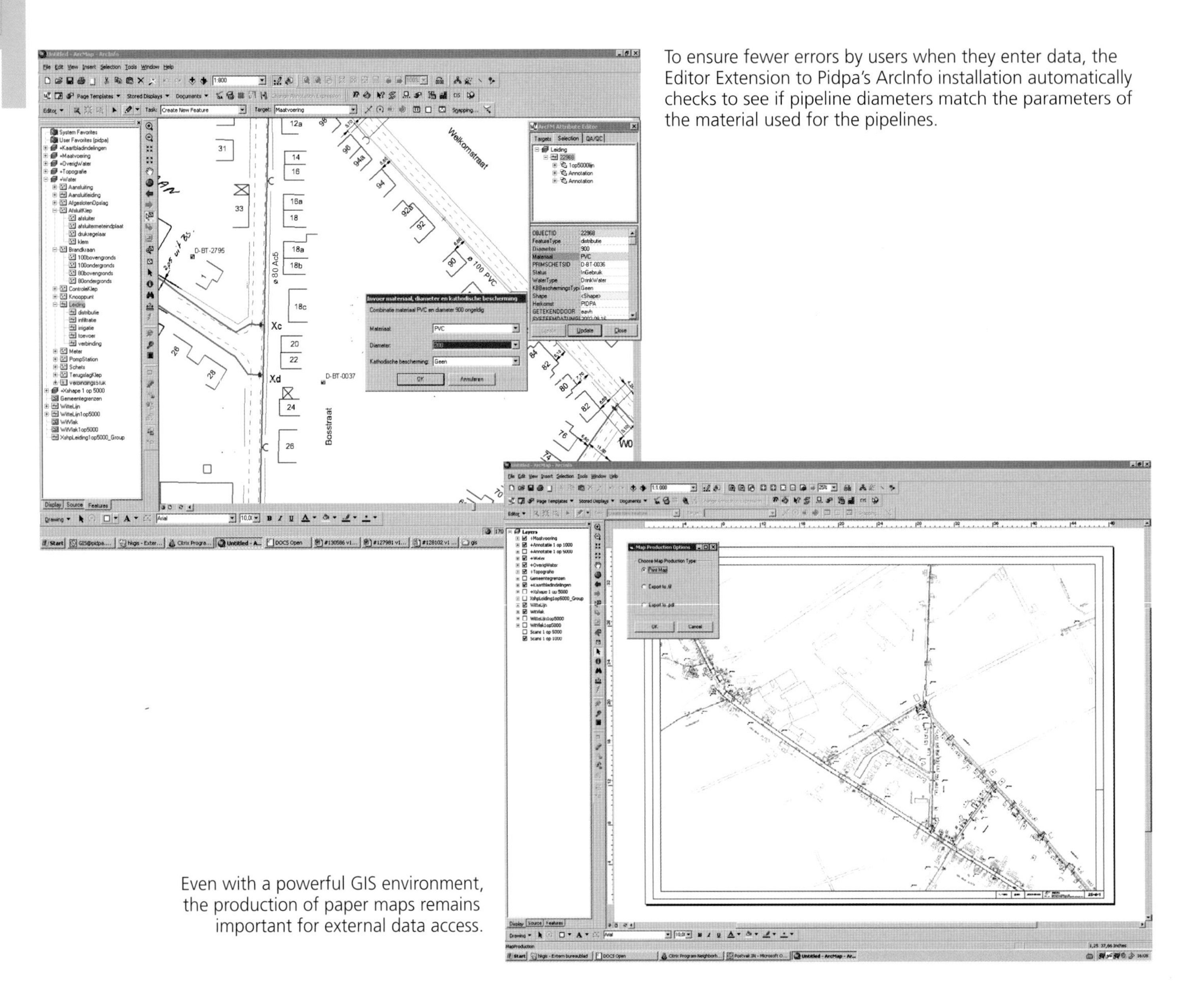

Even with a powerful GIS environment, the production of paper maps remains important for external data access.

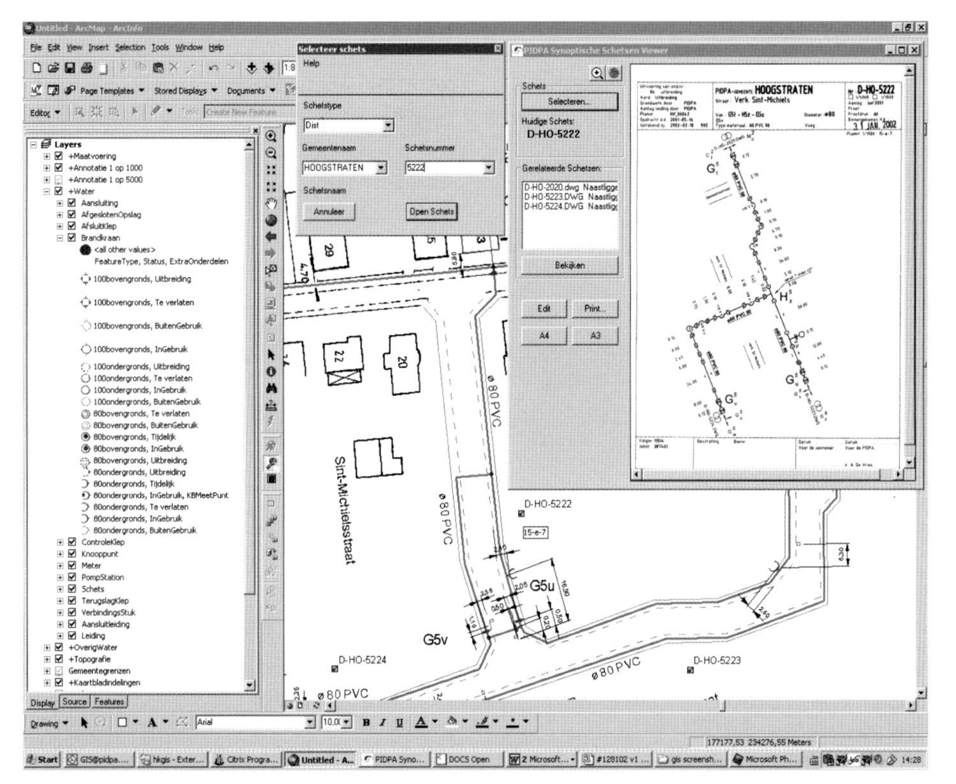

Detailed synoptical drawings provide detailed historical information on the water network structure.

Additional tools

Because of Pidpa's commitment to developing a system that can be deployed throughout Europe, the use of an open system architecture was essential. The new system is based on ArcGIS 8.3 and on ArcFM™ 8.3, a software package from ESRI Business Partner Miner & Miner. Because Pidpa's GIS was to be a turnkey system, some specific tools dedicated to managing a water distribution network were created by ESRI in California and ESRI Nederland B.V.

Map printing

Pidpa has been using paper maps for many years, and managers knew it would be unreasonable to expect employees to adapt to a totally paperless environment almost overnight. To deal with this, a map production tool was created to enhance the cartographic capabilities of ArcGIS. With this tool, users can choose a scale of 1:1000 or 1:5000 and the map will be printed using a standard template with a logo and a frame around the map. A user-

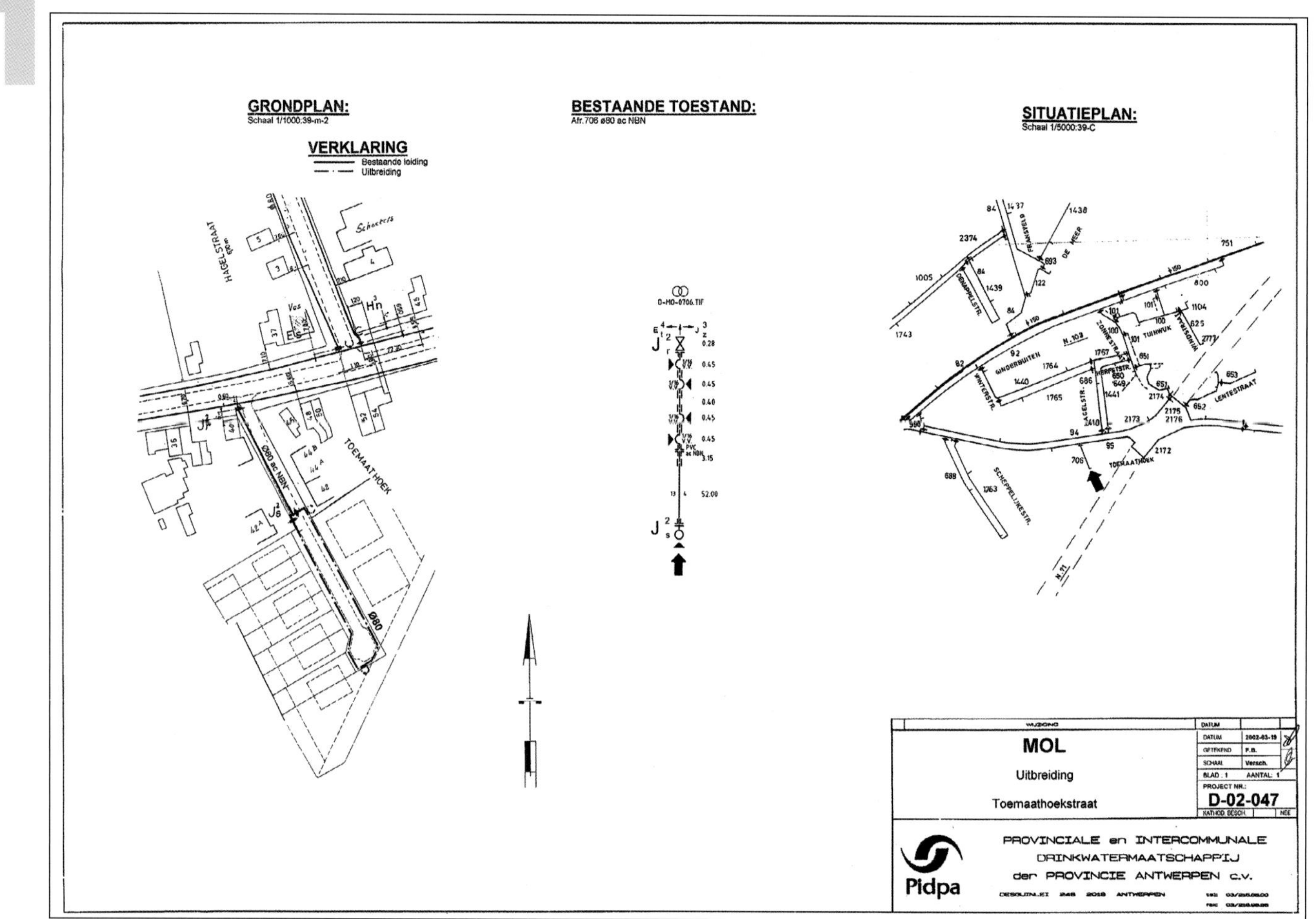

New construction work is now prepared directly in ArcInfo. This results in major time savings because only one drawing needs to be made.

About fifteen editors actively use GIS to maintain the network data on a daily basis.

supplied title, the current date and a map number are also included.

Editor Extension tool

The Editor Extension carries out rule-based checking to ensure that all data entered into the database is as accurate as possible. For example, on entering details of a new water main, the Editor Extension will make sure that the diameter of the main and its material are a valid, real-world combination. Another feature of the Editor Extension is the automatic rotation of valves, hydrants or other network parts to ensure that they're oriented in the right direction. Connectivity is also enforced while drawing.

Synoptic Viewer tool

The Synoptic Viewer, a custom ArcMap™ tool created by ESRI Nederland B.V., allows a user to call up the previously mentioned synoptic drawings by municipality name, by water-main type, or by drawing number. A DWG representation will be displayed if available; if not, a raster image will display. All the drawings are linked, so that when a particular drawing is shown, a list of associated drawings will display. By clicking on one of the listed drawing numbers, a user can browse through the entire library.

User benefits

When a new water main is planned, one of the first tasks to be done is the generation of a design map so the contractor or Pidpa personnel have all the information they need to carry out the work. These design maps can include a portion of a detailed 1:1000-scale map, an extract of an overview map at 1:5000 scale and one or more A4-size synoptic drawings.

Paper maps in a box in a vehicle are a common sight within many European companies. At Pidpa maps have been digitised.

Mobile devices are being rolled out to Pidpa field crews. This allows GIS and corporate SAP data to be viewed in the field.

In the past, these design maps were created manually on paper and required much copying and pasting of extracts from other existing paper documents. Once the work was done, the maps had to be redrawn to reflect the changes. Today the process has been streamlined with GIS. In the planning stage, the geodatabase is edited to reflect the work that needs to be done—whether repairing, constructing, removing, or some similar task. A document based on a predefined ArcFM page template is generated. When the work is completed and field remarks have been processed, the status of the affected network items is simply changed to 'in use'.

Pidpa has a large number of crews in the field who rely on geographic information about the water distribution network. In the past, a field crew had only a box of paper 1:5000-scale overview maps in the vehicle to work with. Even with all this paper, critical information could still be missed, jeopardizing the crew's ability to respond efficiently to problems.

Today, a pilot project has started to provide the field crews with access to GIS data. Using a laptop equipped with a mobile printer, any map can be retrieved and printed at a scale of 1:1000 or 1:5000; all the 90 000 detailed synoptic drawings are also available. Initially, these maps and diagrams will be stored on the laptop's hard drive, but eventually the system will be expanded with a mobile-phone GPRS link that will allow the crew to use live GIS data using ArcIMS.

The system allows design maps to be created in GIS as a basis for accurate field work. Afterward the system is updated to reflect the work carried out.

The GeoLink ArcIMS viewer offers fast access to GIS features but also allows users to retrieve linked corporate data via interfaces to SAP and other proprietary databases.

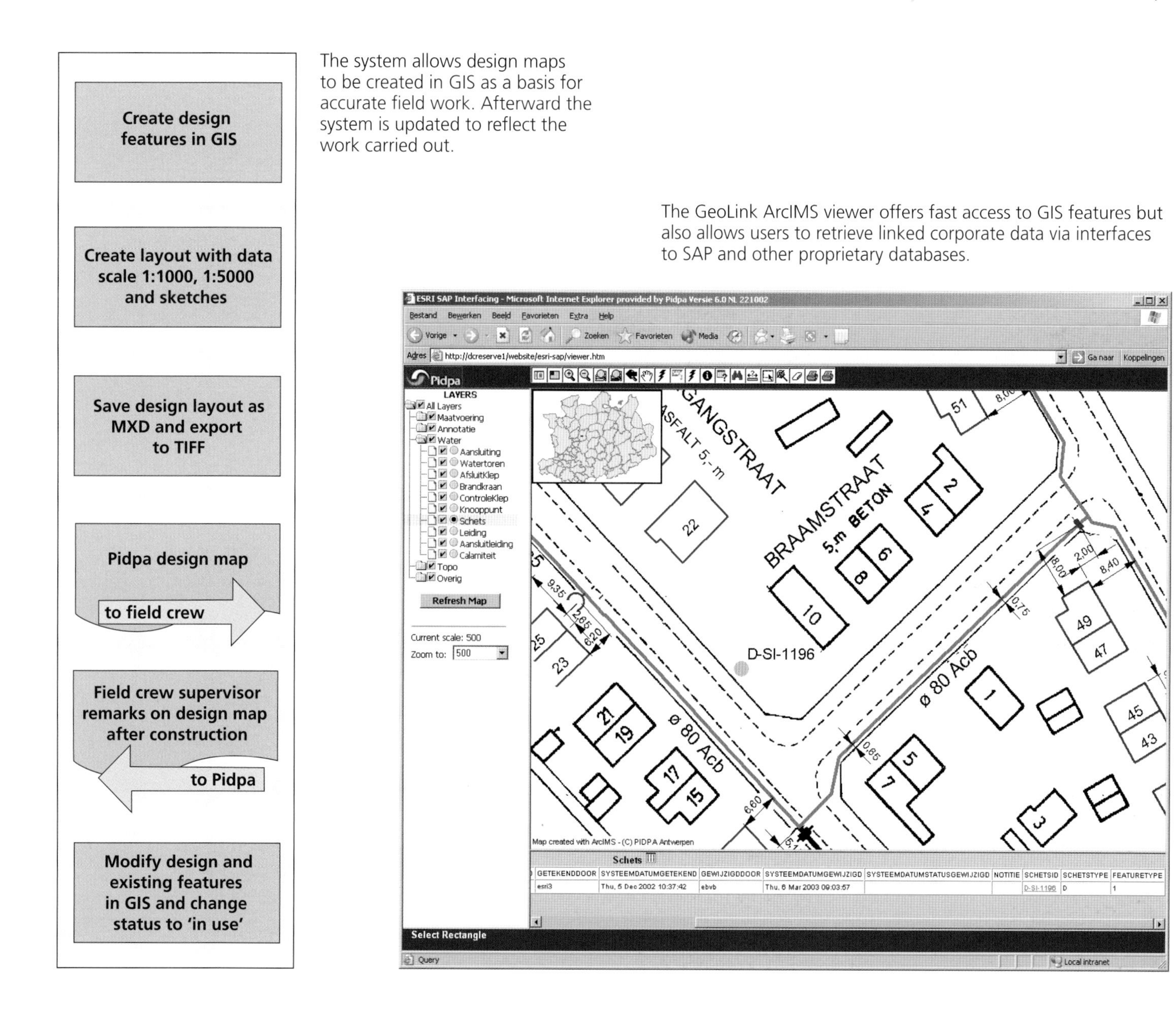

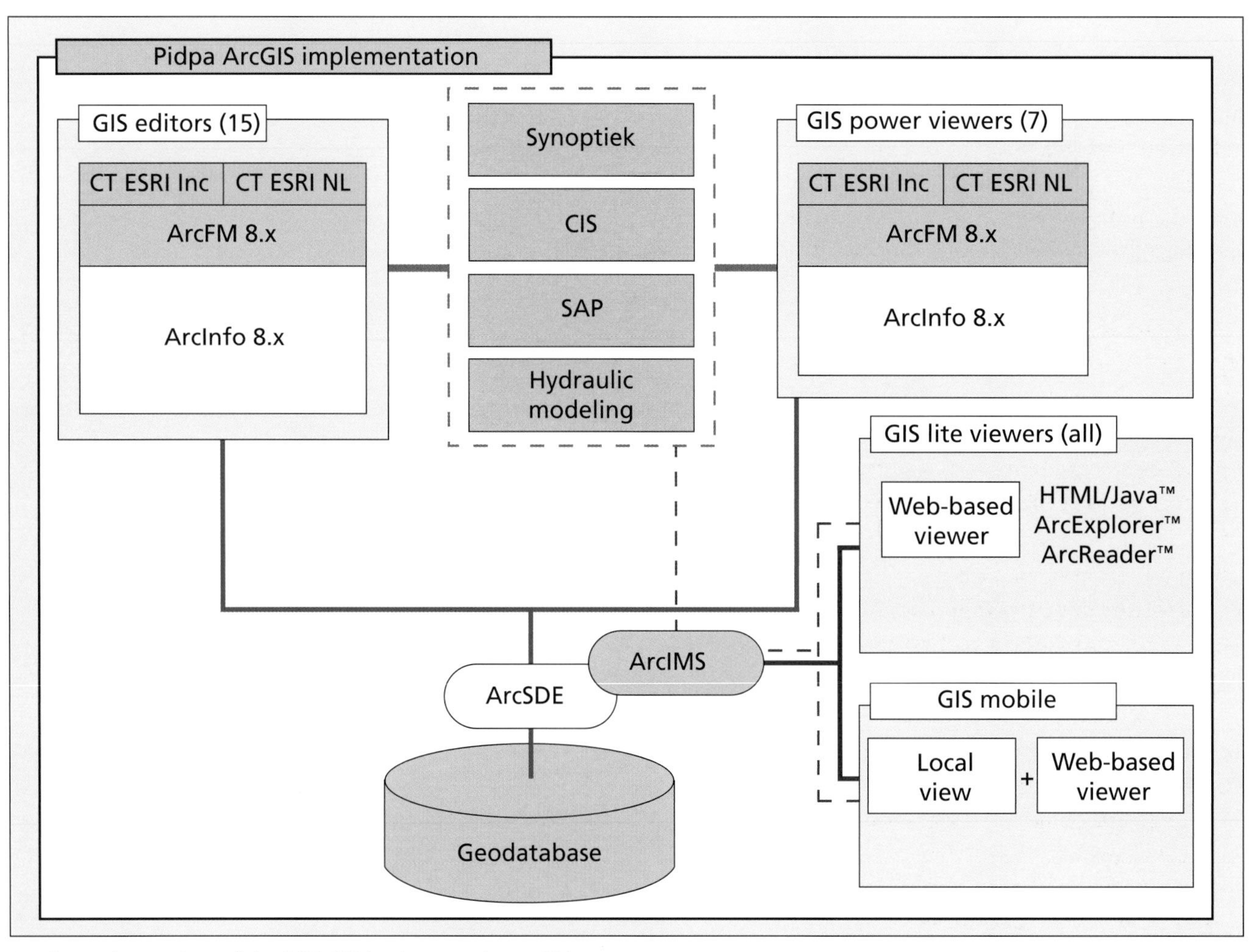

A schematic overview of the ESRI GIS implementation at Pidpa

The road ahead

Future plans involve integrating GIS with many of the other software systems in use throughout Pidpa.

To date, the main link is to SAP, which allows work order, project and location information to be retrieved directly from the ArcIMS environment. Additionally, there is a link to the customer information system (CIS) on an IBM® mainframe, which provides easy access to a vast amount of customer information, as well as a link to the supervisory control and data acquisition (SCADA) system, which allows a user to view real-time project status information from the GIS with only a mouseclick. By sharing data among these environments and optimising workflow across the enterprise, the system will let users get to the data they need more easily. GIS can be the glue to make all this happen for any utility company.

As the use of vector data grows in GIS, more analytical possibilities become available, which in turn supports management in decision making. One goal is to use GIS data with hydraulic modeling software and to have planners run simulations based on complete GIS vector data. Extensive tracing functionality will be implemented once Pidpa's water distribution network data is completely converted to vector format.

Another major challenge for the future is to further integrate field crew work processes with the workflow system in the office. Instead of static information, Pidpa aims to equip field crews with mobile devices that allow easy access to accurate and up-to-date GIS and SAP information.

Hardware

A thin-client configuration is used to enable 15 concurrent editing and seven viewing licenses on a variety of desktop PCs acting as client workstations, linked to both standard and large-format printers.

The application server is a Dell® four-way iP Xeon 700 machine with 1.5 GB RAM for the viewing clients, and a Dell two-way iP 2.8-GHz machine with 4 GB RAM for the editing clients. The database server is a four-way iP Xeon 700 server with 2.5 GB RAM that connects directly to a central storage network (SAN) that stores the database and raster data.

The ArcIMS server is a Dell two-way iP 2.8-GHz machine with 4 GB RAM.

Software

ArcGIS 8.3 and ArcFM 8.3, from Miner & Miner, plus additional custom tools created by ESRI and ESRI Nederland B.V.

Data

The GIS uses one centrally located Microsoft SQL Server 2000 database with ArcSDE 8.3. This database contains all water distribution network data as well as the topographic data. Information about synoptic sketches is stored in a separate SQL Server database.

Connections are made to a DB2® database on an IBM mainframe for customer information data, and an SQL Server 2000 database for SAP data.

Acknowledgments

Thanks to Bart Reynaert, Rene Horemans and Patrick Vercruyssen of Pidpa.

7 What's in your backyard?

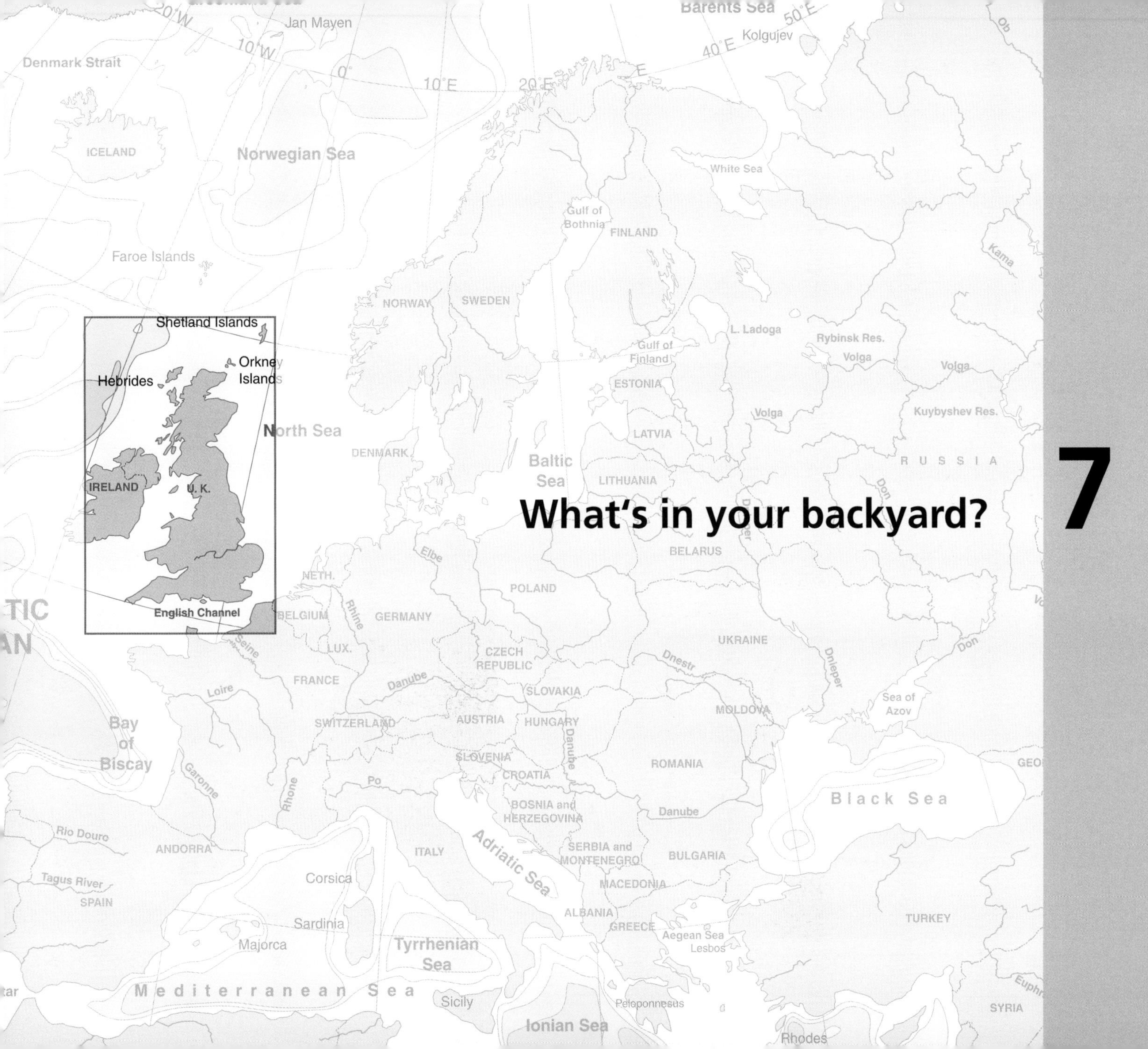

People seldom consider the extent to which their lives depend on control of the water around them: the quality of the water they drink must be kept to the highest standards; the water in rivers and streams near their homes must be kept safely contained within its channels; the water in the oceans where they swim must be kept clean.

In the United Kingdom, the job of all this control and monitoring is that of the Environment Agency in England and Wales. To reduce the gap between public perception of water safety and sanitation, and reality—as well as to simply give people more knowledge about the environment they live in—the EA has put together an information-rich Web site. Called 'What's in Your Backyard', the site has been live for more than four years, with steadily increasing popularity and rising visibility. In early 2004, the site was routinely recording more than 400 000 page hits per month.

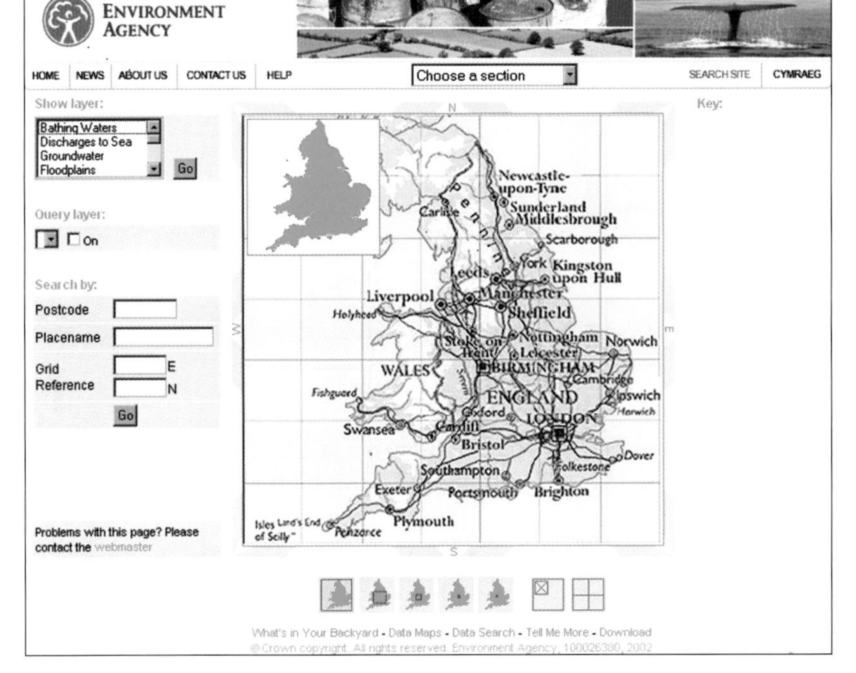

When users of the Web site select the Data Maps option, they are able to pinpoint sites of interest by postal code, place-name or national grid reference. They can then view environmental data sets from a national or local perspective.

Early users tended to be specialists, particularly knowledgeable about the environment and well-versed in interpreting environmental statistics. But now, the site has a much broader user base, from academics to Web-conscious citizens to computer novices.

A major aim of the Web site was to give the public information with which to make informed decisions; the breadth of the current user base is indicative of the success of that goal. And to make environmental data even more accessible, the Environment Agency is constantly appraising the Web front end with a view to making it even more intuitive and user friendly.

Currently, the site provides access to 13 major databases, and more are planned. Most of the data is map based and generated on the fly from a MapObjects®–MapObjects Internet Map Server back end.

Typically, a user will enter a postal code application, place-name or national grid reference and the browser will then show a map of that area. These maps, which are based on raster data from the British Ordnance Survey at scales of 1:10 000, 1:50 000 and 1:250 000, can be zoomed and panned at will. Selecting a query layer brings detailed textual, tabular and graphical information linked to icons on the map. And because the Web site is designed to be accessible and

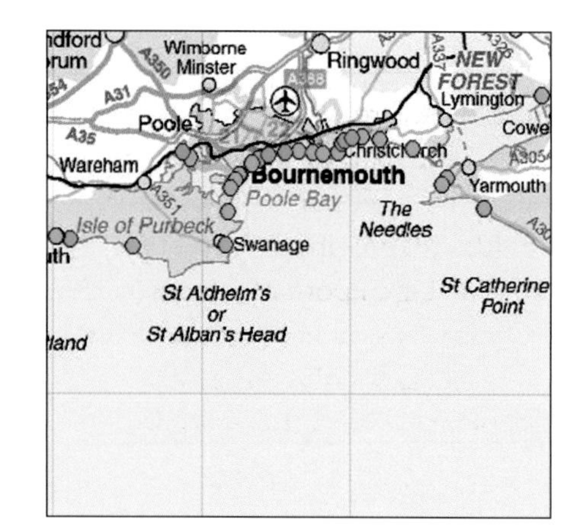

In this example, the Bathing Water data set for the southeast coastline of Dorset has been selected and displayed on a map showing the location of sampling sites and a colour-coded assessment of bathing water quality.

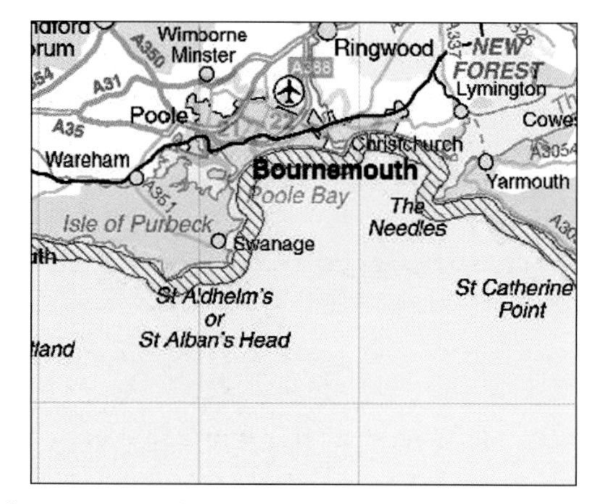

The Discharges to Sea data set can be displayed as a map of the coastal zones that are monitored for the OSPAR Convention. If the Query layer is switched on, clicking on a zone will give access to data on loads and sources of polluting substances discharged into the zone.

Using Map Search, the Flood Plains data set shows the extent of natural river and coastal floodplains.

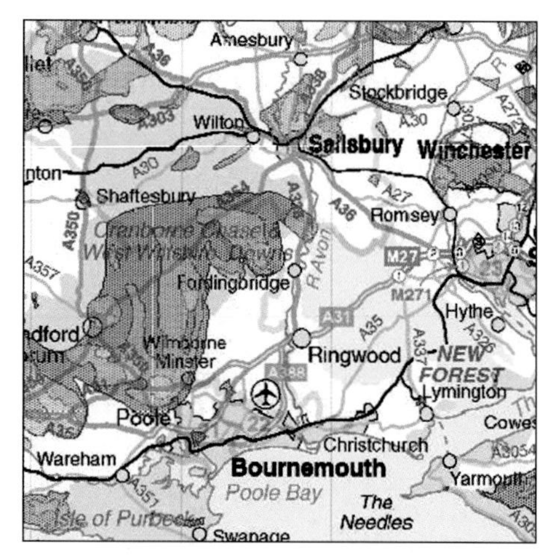

On the map, the Groundwater Source Protection Zones data set provides an indication of the risk to groundwater of polluting activities and accidental releases of pollutants.

understood by ordinary citizens, additional pages—titled Tell Me More—are available for each of the data sets. These provide background information on the data set, how the data is obtained and, most importantly, how to interpret it.

Forewarned is forearmed

Severe flooding affected many parts of the United Kingdom in the late 1990s and in the early part of this century. It's not surprising, especially in view of massive media coverage, that flooding danger is now a matter of considerable public concern. Thus, the Indicative Floodplain maps on the site are one of the most frequently accessed parts of What's in Your Backyard.

Often there's little that can be done to prevent flooding, but householders can prepare. The Environment Agency Web site provides helpful information for doing so. Making up a flood kit, storing important papers (such as insurance policies) in a safe place, compiling a list of important phone numbers, and buying sandbags are all things that those living on a floodplain are encouraged to do.

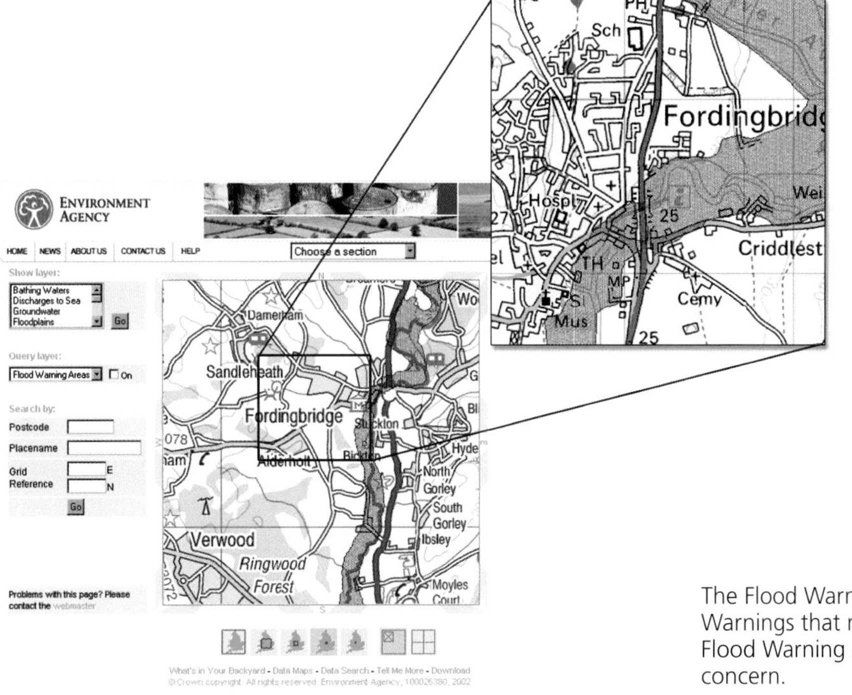

The Flood Warning Areas data set may be examined on the map. Warnings that may be in force can be viewed by switching on the Flood Warning Areas Query Layer and then clicking on the area of concern.

Keeping the waters at bay

The Environment Agency also offers constituents two ways of assessing actual flood danger. Besides the the Indicative Floodplain maps—which contain static data updated annually—there are the Flood Warning maps, which are integrated with dynamic data, providing immediate, near-real-time information on any current flooding threat level. These maps make compelling reading for those living in flood-prone areas.

In specially designated Flood Warning Areas—those populated areas that are prone to flooding, and where it's feasible to provide advanced warning—local Environment Agency employees monitor the situation continually. With an expert knowledge of the local area, and up-to-the-minute information on rainfall, groundwater and river conditions, one of four escalating warning codes is assigned to an area. These codes—All Clear, Flood Watch, Flood Warning and Extreme Flood Warning—indicate the severity of flooding and the level of danger. For those who have heeded the Environment Agency's recommendation to create a flood plan, receiving a flood warning is the signal to put that plan into operation.

River water quality

In few places is the effect of pollution more obvious than in local rivers and canals. Years of industrial discharge have affected the health and diversity of ecosystems, the suitability of drinking water supplies, and recreational activities. In 1990, 76 sites in the United Kingdom had high levels of very toxic pollutants such as mercury and the pesticide dieldrin.

In recent years, river quality standards have been established in England and Wales, and measures have been put in place to ensure that rivers meet those standards. What's in Your Backyard includes data sets on these standards and information about how well individual rivers meet them. Included in this section is data on river chemistry, biology, nutrients and even—for those areas that attract tourists—aesthetics. In addition to the current status, users can view trend graphs to see how the situation has changed over the years. In 11 years of monitoring, those 76 sites with levels of mercury and dieldrin had fallen to only two.

More detailed trends can also be viewed, thanks to automated monitoring stations that record data every hour. Used mainly for research and education, these pages allow students to learn about the environment. Students can see, for example, how dissolved oxygen levels—an indicator of water quality—are affected by plant photosynthesis and in turn, by time of day.

Keeping a watch on discharge

Knowing how much chemical contaminant there is in your local river is one thing; knowing where the pollution comes from is another. To help the public keep an eye on discharges from nearby industrial plants, the Environment Agency has provided a pollution inventory on the Web site that measures discharges into rivers, as well as airborne pollution.

Many large industrial sites are required to report annually the quantities and varieties of potentially harmful substances they have released into the environment. The types

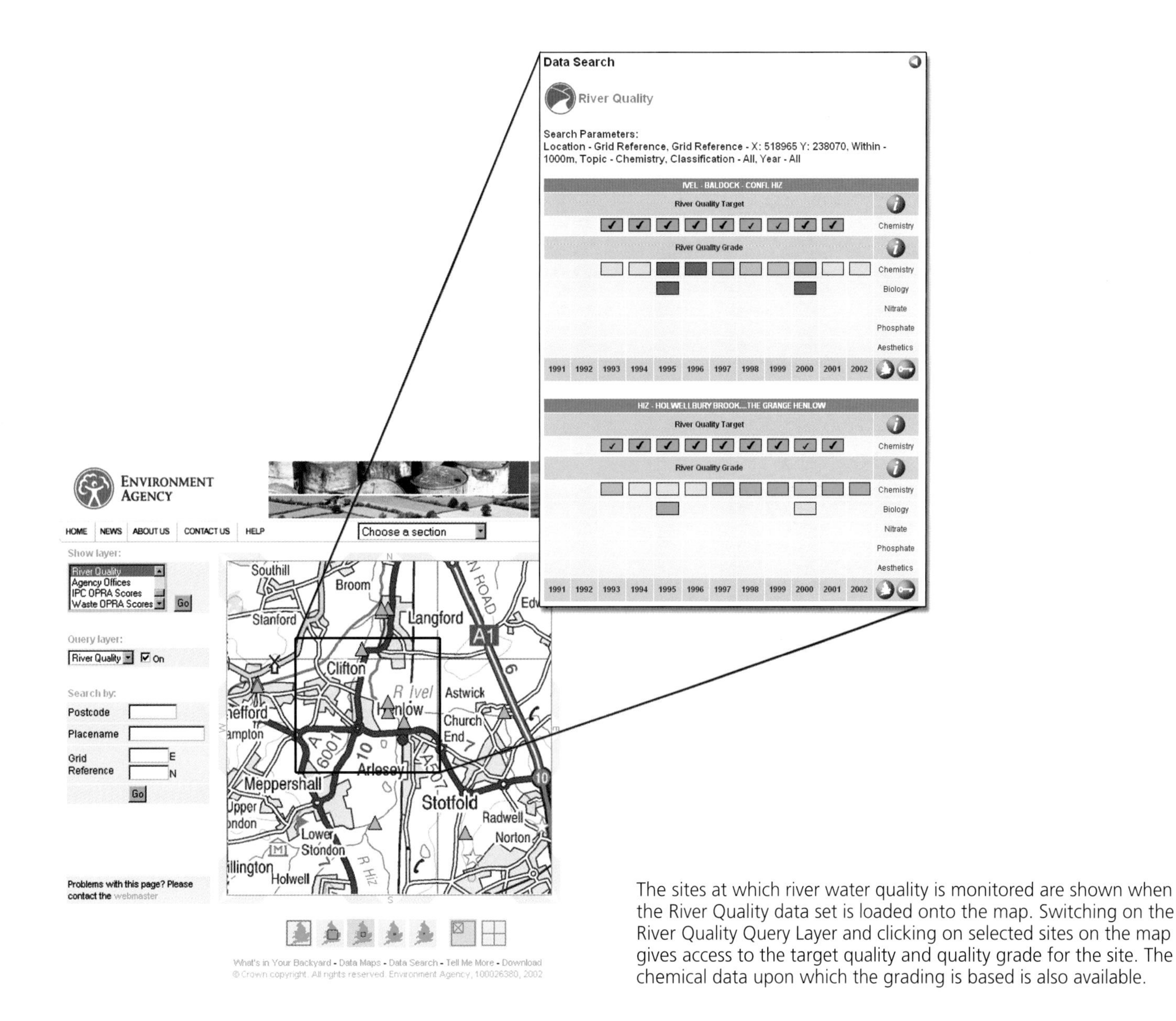

The sites at which river water quality is monitored are shown when the River Quality data set is loaded onto the map. Switching on the River Quality Query Layer and clicking on selected sites on the map gives access to the target quality and quality grade for the site. The chemical data upon which the grading is based is also available.

The Discharges to Sea data set shows the monitoring zones for the OSPAR Convention, a European agreement to protect the quality of the northeast Atlantic. With the Discharges to Sea Query Layer switched on, clicking on a selected coastal monitoring zone will show the annual loads of polluting substances released into that zone. When the appropriate button for a selected substance is clicked, more detailed information on type of source and load estimate is given.

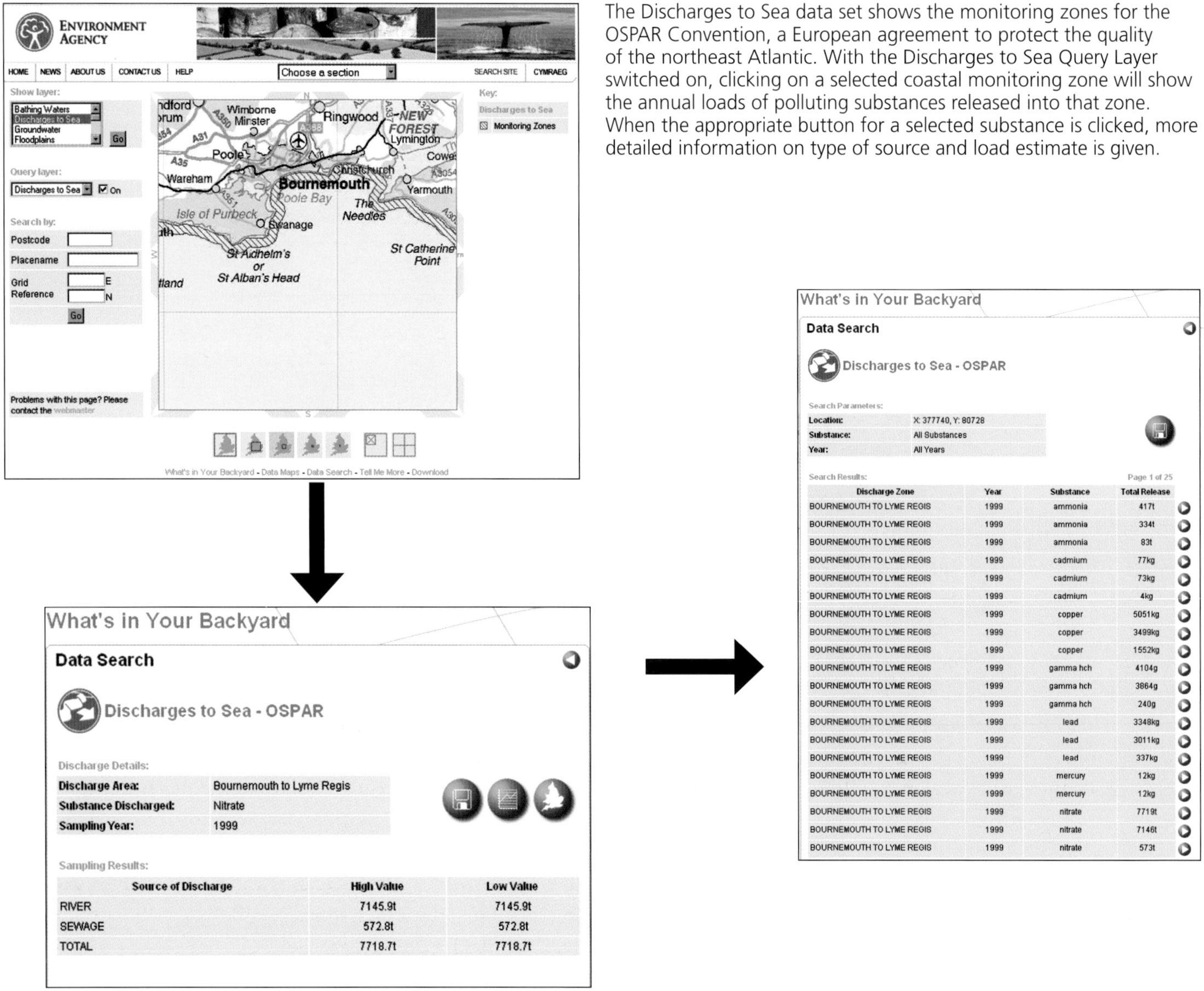
Environment Agency
HOME NEWS ABOUT US CONTACT US HELP
Choose a section
SEARCH SITE CYMRAEG
Show layer:
Bathing Waters
Discharges to Sea
Groundwater
Floodplains
Go
Query layer:
Discharges to Sea
On
Search by:
Postcode
Placename
Grid Reference
E
N
Go
Problems with this page? Please contact the webmaster
Key:
Discharges to Sea
Monitoring Zones
Wimborne Minster
Ringwood
NEW FOREST
Lymington
Poole
Wareham
Christchurch
Bournemouth
Poole Bay
Yarmouth
Isle of Purbeck
Swanage
The Needles
St Catherine Point
St Adhelm's or St Alban's Head
What's in Your Backyard - Data Maps - Data Search - Tell Me More - Download
What's in Your Backyard
Data Search
Discharges to Sea - OSPAR
Discharge Details:
Discharge Area: Bournemouth to Lyme Regis
Substance Discharged: Nitrate
Sampling Year: 1999
Sampling Results:
Source of Discharge | High Value | Low Value
RIVER | 7145.9t | 7145.9t
SEWAGE | 572.8t | 572.8t
TOTAL | 7718.7t | 7718.7t
What's in Your Backyard
Data Search
Discharges to Sea - OSPAR
Search Parameters:
Location: X: 377740, Y: 80728
Substance: All Substances
Year: All Years
Search Results:
Page 1 of 25
Discharge Zone | Year | Substance | Total Release
BOURNEMOUTH TO LYME REGIS | 1999 | ammonia | 417t
BOURNEMOUTH TO LYME REGIS | 1999 | ammonia | 334t
BOURNEMOUTH TO LYME REGIS | 1999 | ammonia | 83t
BOURNEMOUTH TO LYME REGIS | 1999 | cadmium | 77kg
BOURNEMOUTH TO LYME REGIS | 1999 | cadmium | 73kg
BOURNEMOUTH TO LYME REGIS | 1999 | cadmium | 4kg
BOURNEMOUTH TO LYME REGIS | 1999 | copper | 5051kg
BOURNEMOUTH TO LYME REGIS | 1999 | copper | 3499kg
BOURNEMOUTH TO LYME REGIS | 1999 | copper | 1552kg
BOURNEMOUTH TO LYME REGIS | 1999 | gamma hch | 4104g
BOURNEMOUTH TO LYME REGIS | 1999 | gamma hch | 3864g
BOURNEMOUTH TO LYME REGIS | 1999 | gamma hch | 240g
BOURNEMOUTH TO LYME REGIS | 1999 | lead | 3348kg
BOURNEMOUTH TO LYME REGIS | 1999 | lead | 3011kg
BOURNEMOUTH TO LYME REGIS | 1999 | lead | 337kg
BOURNEMOUTH TO LYME REGIS | 1999 | mercury | 12kg
BOURNEMOUTH TO LYME REGIS | 1999 | mercury | 12kg
BOURNEMOUTH TO LYME REGIS | 1999 | nitrate | 7719t
BOURNEMOUTH TO LYME REGIS | 1999 | nitrate | 7146t
BOURNEMOUTH TO LYME REGIS | 1999 | nitrate | 573t

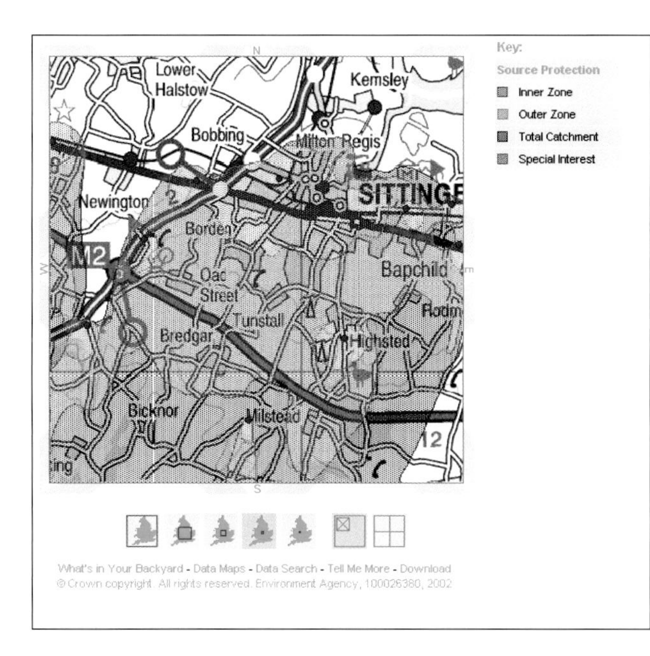

The Groundwater data set gives information on zones around groundwater abstraction points. The zones are based, for example, on estimates of the catchment area for the water supply source and travel time of contaminants in groundwater.

of sites covered are those that, according to legislation, have the greatest potential to cause damage. Maps on What's in Your Backyard show those locations that are required to report on their emissions. By clicking an icon, users can read the most recently completed pollutant-release report. But the pollution inventory isn't just intended to provide information. The Agency expects that it will also be a vehicle for change; river water quality trends indicate that this may be happening.

Protecting water supplies

Groundwater provides more than a third of the United Kingdom's drinking water supply and it also plays an important role in maintaining the flow of many rivers. Its protection, therefore, is something the Environment Agency takes very seriously.

Groundwater is at considerable threat from accidental releases of pollutants. But unlike spillage into a river, it's not immediately obvious to the naked eye that a spill has occurred and that it could affect drinking water supplies. Groundwater pollution can also be spread over a larger geographic area, and even if discovered, it's much harder to remove. For these reasons, the Environment Agency has designated so-called Source Protection Zones that are associated with wells, boreholes and springs. These zones indicate how quickly chemicals spilled onto the ground will pollute these underground water sources.

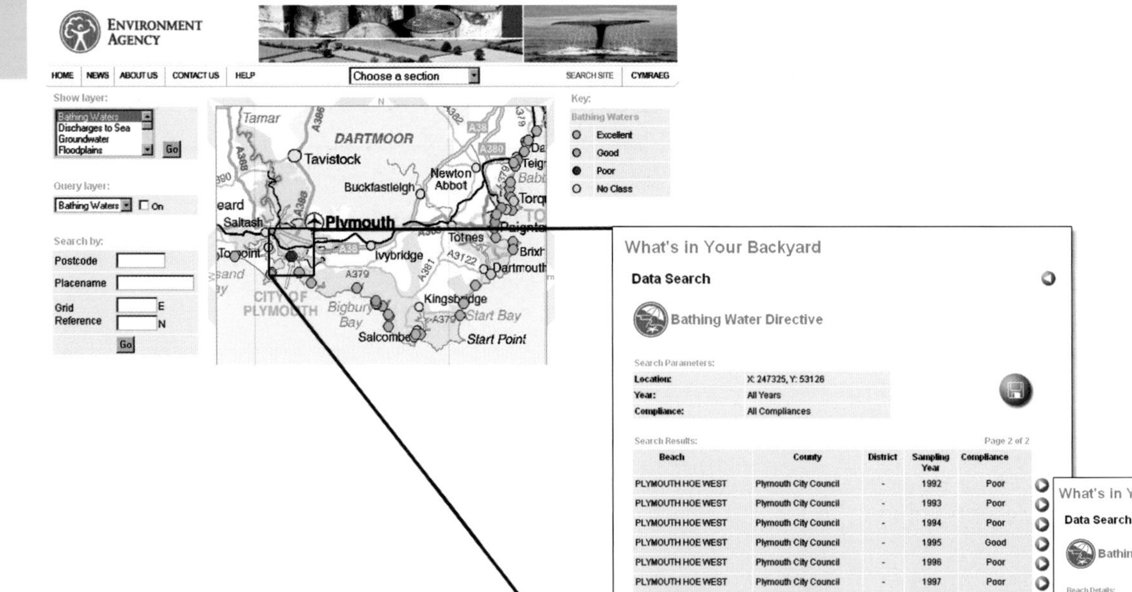

What's in Your Backyard

Data Search

Bathing Water Directive

Search Parameters:

Location:	X: 247325, Y: 53126
Year:	All Years
Compliance:	All Compliances

Search Results: Page 2 of 2

Beach	County	District	Sampling Year	Compliance
PLYMOUTH HOE WEST	Plymouth City Council	-	1992	Poor
PLYMOUTH HOE WEST	Plymouth City Council	-	1993	Poor
PLYMOUTH HOE WEST	Plymouth City Council	-	1994	Poor
PLYMOUTH HOE WEST	Plymouth City Council	-	1995	Good
PLYMOUTH HOE WEST	Plymouth City Council	-	1996	Poor
PLYMOUTH HOE WEST	Plymouth City Council	-	1997	Poor
PLYMOUTH HOE WEST	Plymouth City Council	-	1998	Poor
PLYMOUTH HOE WEST	Plymouth City Council	-	1999	Good
PLYMOUTH HOE WEST	Plymouth City Council	-	2000	Good
PLYMOUTH HOE WEST	Plymouth City Council	-	2001	Good
PLYMOUTH HOE WEST	Plymouth City Council	-	2002	Poor
PLYMOUTH HOE WEST	Plymouth City Council	-	2003	Not Classified

What's in Your Backyard

Data Search

Bathing Water Directive

Beach Details:

Beach Name:	Plymouth Hoe West
County Authority:	Plymouth City Council
Sampling Year:	2002

Sampling Results:

Date	Time	Total Coliforms (colonies / 100ml)	Faecal Coliforms (colonies / 100ml)	Faecal Streptococci (colonies / 100ml)	Salmonella	Enterovirus (plaque forming units / 10 ltr)	Water Quality
01/05/02	13:05	36	27	10	-	-	Excellent
13/05/02	11:00	9600	150	54	-	-	Good
21/05/02	12:10	140	81	27	-	-	Excellent
24/05/02	10:55	5040	2700	530	-	-	Poor
11/06/02	10:35	21000	20000	180	-	-	Poor
20/06/02	12:45	36	10	10	-	-	Excellent
27/06/02	12:35	18	10	10	-	-	Excellent
05/07/02	14:25	120	36	10	-	-	Excellent
11/07/02	12:08	10	10	10	-	-	Excellent
17/07/02	14:25	240	36	81	0	-	Excellent
25/07/02	12:23	27	18	10	-	-	Excellent
29/07/02	12:20	10	10	10	-	1	Excellent
02/08/02	14:00	18	10	10	-	-	Excellent
05/08/02	13:30	132	162	27	0	-	Good
12/08/02	15:35	1840	1818	171	-	-	Good
25/08/02	11:49	41	10	10	-	-	Excellent
29/08/02	11:40	120	27	10	-	2	Excellent
04/09/02	13:35	45	10	10	-	-	Excellent
10/09/02	10:40	10	18	18	-	-	Excellent
23/09/02	12:05	103	10	18	-	-	Excellent

The Data Maps facility in What's in Your Backyard shows where environmental data is collected. This example shows the sites that are monitored for compliance with the European Bathing Water Directive. The user can then obtain information on the annual assessments of compliance and access the underlying data.

Information on Source Protection Zones is used by those regulatory bodies responsible for new industrial or agricultural development; some activities of new developments would be designated as unacceptable use in these zones. Source Protection Zone information is also used by government emergency services to gauge the potential effect of accidental discharge—for example, a traffic collision involving a fuel tanker.

Clean beaches for all

The quality of bathing water can be an emotional issue because pollutants in the sea can have immediate, devastating consequences not only for public health but also for the tourist industry. The cleanliness of bathing waters is controlled by regulations known as the Bathing Waters Directive, and information on how well beaches comply with that directive is available online on the What's in Your Backyard Web site.

Throughout the bathing season, Environment Agency scientists sample the water at each beach, and samples are analyzed for bacteria indicative of sewage pollution. This analysis, which may take up to two weeks to complete, is then made available on the Web site. Users can read the analysis by clicking on the map icon associated with a particular stretch of bathing water.

The benefits of this kind of heightened scrutiny are plain to see. From 2000 to 2002, the percentage of bathing waters consistently falling below standard was less than 0,5 percent. That is a dramatic reduction: during a similar two-year period a decade earlier, the percentage of substandard water stood at 13 percent.

Speaking in 2002, to celebrate the addition of 14 beaches to the safe-beach category, which in turn allowed the awarding of the prestigious Blue Flag award, Environment Minister Michael Meacher said, 'The British seaside has never been as safe and clean as it is today'.

Hardware

Development: Compaq DL360 desktop PC (866-MHz Pentium III, 512 MB RAM, 36-GB disk, Windows 2000)
Web hosting (using n+1 redundancy to achieve 99.97 percent uptime):
SQL Servers: Compaq DL3600 (Windows NT)
Map servers (MapObjects and MapObjects Internet Map Server): Compaq DL3600 (Windows NT)

Software

MapObjects 2 and MapObjects Internet Map Server 2.
ArcView 3.1 is used for day-to-day data editing.

Data

The following are Environment Agency-generated relational databases running on Microsoft SQL Server 7:
Bathing Waters Directive
OSPAR (Discharges to Sea)
Source Protection Zones
Indicative Floodplain Maps
Flood Warning Areas
Landfill Sites
Pollution Inventory
OPRA (the Agency's rating of the pollution hazards and the operation of factories)
General Quality Assessment (river water quality)
River Quality Objectives
Agency Offices
Catchment Abstraction Management Strategies
Environment Agency Public Face Boundaries

Third-party data

Local Authority Boundaries
Ordnance Survey raster maps at 1:10 000, 1:50 000 and 1:250 000

Web site

www.environment-agency.gov.uk/wiyby

Acknowledgments

Thanks to Bob Huggins of the Environment Agency.

8 Integrating maps, integrating nations

More than two dozen water boards in the Netherlands each cover an area of between 200 and 3000 square kilometres, each with its own system for creating and storing maps and reports for its unique set of geographic data.

Such diversity is not a problem when each board is working with its own discrete data within its own discrete region. But problems can arise when the water boards must work together to build maps covering a wider area. Under the European Water Framework Directive, substantive new legislation from the European Union, this was the task they faced.

This chapter shows how GIS helps water authorities work together to deliver maps and other relevant geographic information on a regular basis.

Three water boards within the Netherlands that collaborated to develop a uniform data set for their waterways

The European Water Framework Directive

The European Water Framework Directive (WFD) sets out how European Union member nations can improve water quality. Historically high pollution levels in large part created the impetus for the directive—pollution that had plagued many of the major rivers in Europe, including the Rhine and the Meuse, for decades. (In the 1970s, the Rhine was often referred to as a "sewer" because of its high pollution levels.) Although water quality has improved since then, there is still considerable room for improvement.

The need for developing more comprehensive European water legislation was first identified by the European Council of Ministers in 1988. The resulting current legislation took effect in December 2000. It states that water is a heritage to be protected, and that clean water systems in clean ecologies are the goal everywhere in Europe. It is a lofty goal, but one that does not have to be reached completely until 2027. To begin the process, an intricate set of rules and schedules—a framework—has been formulated. The WFD is that framework.

One of the earliest of the preliminary steps is the creation of a common, universally accepted definition of the water systems involved. This means defining common terms such as *lake* or *canal.* All European member states and prospective members of the Union should agree on, and use, the same definitions.

A further step in the process is an accurate assessment of the current state of each water system, based on criteria such as biological and chemical conditions, health of aquatic life, and concentrations of pollutants. Quality targets must be formulated and a monitoring programme implemented to ensure that the various measures taken do in fact result in improved water quality. The recommendations for monitoring are different for different parameters: for example, the WFD dictates that the most toxic pollutants be measured at least 12 times a year.

River basins

To meet the goals of the WFD, water managers in the Netherlands have had to intensify their efforts to improve water quality. However, the WFD methodology posed unique problems for them, for example, that water-quality reports must evaluate an entire river basin. But in the Netherlands, no river basin is completely within the country's boundaries; the four that exist cross international boundaries. Data from the various water boards that regulate different portions of the four river basins must be integrated, so maps can be created that encompass an entire river basin, or at least cohesive portions of it. The reports and maps at the national and international level are created from reports of smaller river basins at local and regional levels from all the member states. National regions are created first, then merged to create transnational river basin regions.

The Rhine, for example, flows through several countries, including both the Netherlands and Germany, so the relevant parts of both national water-quality reports must be combined to provide a final report covering the entire Rhine river basin, which covers 224 000 square kilometres.

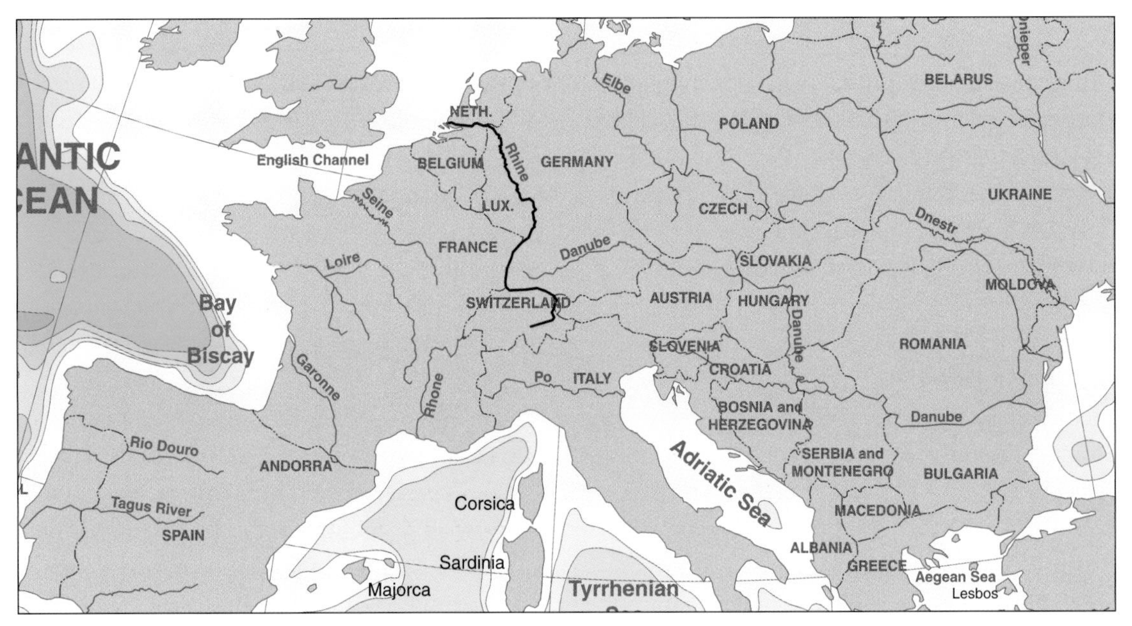

In the Netherlands, no river basin is completely within the country's boundaries, such as the Rhine that also flows through Germany.

Meeting strict requirements

Required river basin data—which must be supplied in the form of digital or hard-copy maps—goes to European Union headquarters in Brussels, on a strict timetable. Member states need to finish the description of the river basins within their national borders by the end of 2004. By the end of 2009, the various national river basin management plans must be integrated to form an international plan.

Linking information

One prerequisite for the efficient achievement of these complex goals is clear agreement on standards, as well as effective communication, among the various water management authorities. In the Netherlands, a new organisation has been founded to aid in this process: the Informatiedesk standaarden Water (IdsW). This organisation inventories, coordinates and manages the existing water management data standards.

The transfer of data among water boards and water-related organisations poses its own problems. One in particular is how different water management organisations can create the integrated maps required by the WFD when they each use different data storage systems.

ESRI Nederland B.V., in cooperation with the Union of Water Boards, explored solutions to this problem using GIS.

Making a test run

ESRI Nederland approached three water boards to take part in testing. The boards, Regge en Dinkel, Rijn en IJssel and Rivierenland, all border each other and all are part of the Rhine river basin. Each, however, uses a different system for geographical data storage and creating maps. Regge en Dinkel uses Intwis, a system derived from ArcGIS; Rijn en IJssel uses the ESRI shapefile data format; and Rivierenland uses GIS-ZES, based on General Electric's Smallworld technology. Each system delivers the same kind of result with similar data but in a different way. So the goal became to see whether the water boards could produce exactly the same type of map, using the same kind of data.

There were two ways they could do so. One was to bring all the data together in one place and then construct the map. This method had inherent disadvantages. Data consolidation by itself would be a huge operation, and one that would have to be repeated over and over whenever the maps had to be updated.

But the ESRI ArcMap product presented an alternative. The Intwis, shapefile and GIS-ZES formats could be converted to the standard Dutch water data export format, known as IMWA. Next, the maps could be linked, using OpenGIS standards.

Advantages

There were several advantages to this approach. The various parties involved, such as water boards, the Ministry of Transport, Public Works and Water Management and the provinces, would not need to transfer their data to other parties, but only to transfer their maps. Only one management environment would be necessary to combine the maps.

Additionally, this approach meant the source data could be left untouched, making it easier to keep the data up to date.

A further advantage is that integrated maps are easier to publish on the Internet—thus advancing another of the WFD's goals, which is to improve public and nongovernmental participation in the process.

Leading the way

The success of this pilot study, involving three water boards each with a different way of doing things, served as an important stepping stone to implementation of the system that is now becoming a reality in Europe. In this system, national and international Web portals are being created that incorporate data integration and mapping services, requiring of local water boards only that they upload their local data or maps to the portal. With such methods, with the help of the Internet and GIS, data integration is paving the way for European integration.

Acknowledgments

Unie van Waterschappen for graphics, Rijkswaterstaat RIZA (Institute for Inland Water Management and Waste Water Treatment) and Rijkswaterstaat RIKZ (National Institute for Coastal and Marine Management).

9 Protecting property and life

The rise in the number of catastrophic floods around the world in the past few years is not merely the result of increased media interest in hydrology, but an accurate reflection of real events. The incidence of flooding is genuinely on the increase due to factors both natural and man-made. Climate change has contributed to increased turbulence in the weather system; upstream deforestation has reduced the natural storage of water; previously uninhabited floodplains have been developed—all have combined to put more people than ever at risk from devastating floods.

Climate change and deforestation are combining to make our planet a much wetter place.

Traditionally, floods have been controlled through the construction of physical barriers such as dams, flood embankments and diversion channels. While these measures have succeeded in reducing the impact of flooding in the protected areas, the problem is often just moved from one place to another. In some cases, the overall result can be worse than it would be had the protective measures not been taken at all. For this reason, many experts are changing their emphasis from flood control to flood management.

In this chapter we look at the flood management approach of DHI Water and Environment of Denmark. Their MIKE 11 flood-modeling system software technology has been effective in two European countries in planning for flooding contingencies and in providing automated flood warnings.

Models

MIKE 11 Flood Watch is a tool for modeling the dynamic processes in open and closed channels, rivers and floodplains. It uses two distinct elements to do its work.

The first element is a hydrological model—a set of mathematical equations that describe the way water is stored before being released into rivers. For example, in rural areas water is stored as snow, as groundwater, in the root zone and in the ground; water passes among these different types of storage before its eventual release into the rivers. The input for this model is precipitation data, while the output is a time series, that is, information on how the volume of water entering the rivers varies over time.

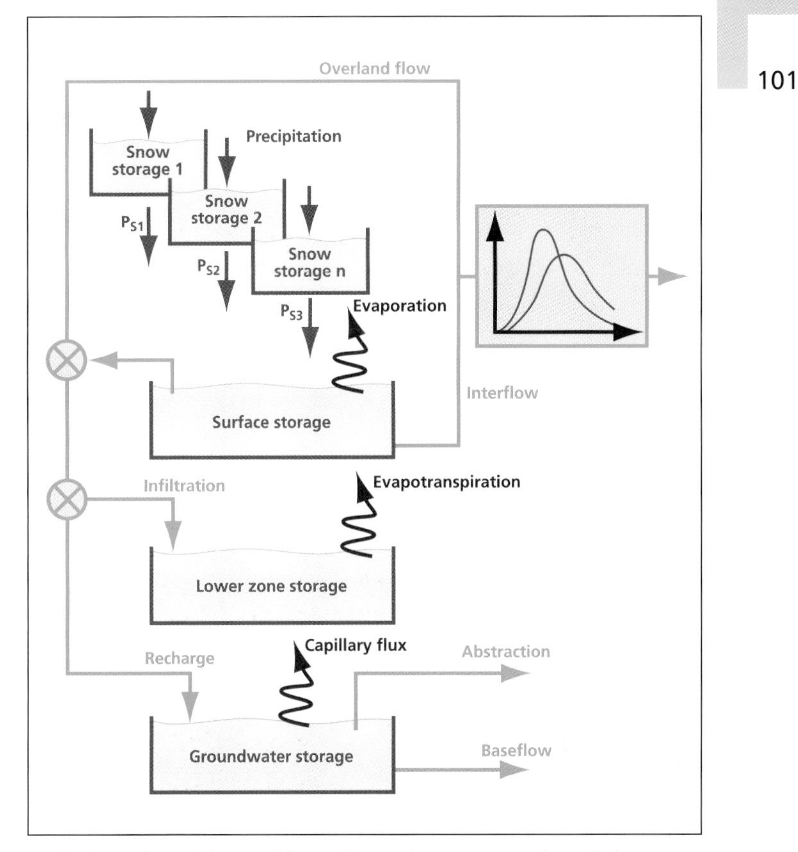

Hydrological models provide a schematic representation of the hydrological cycle and the impact of increased rainfall on the runoff from catchment areas.

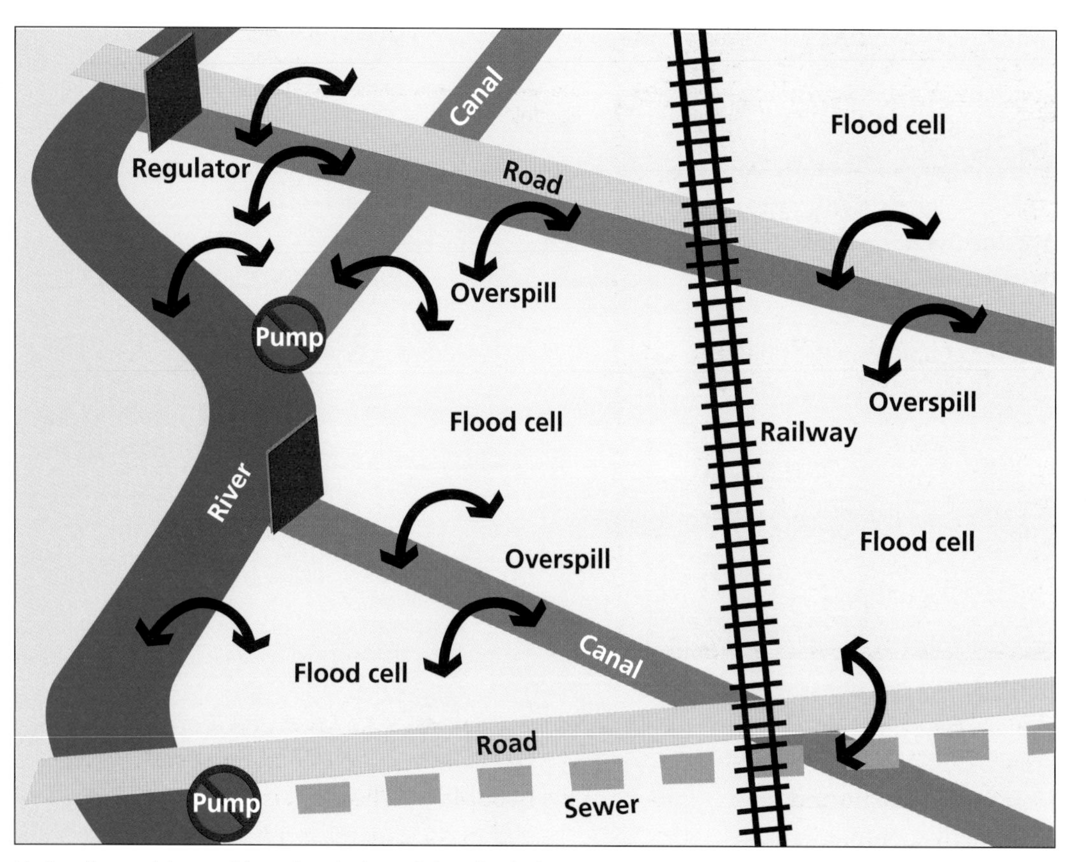

Hydraulic models provide a description of the physical conditions and interactions among a variety of water sources, as well as describing the transfer of floodwater among rivers, canals and floodplain areas.

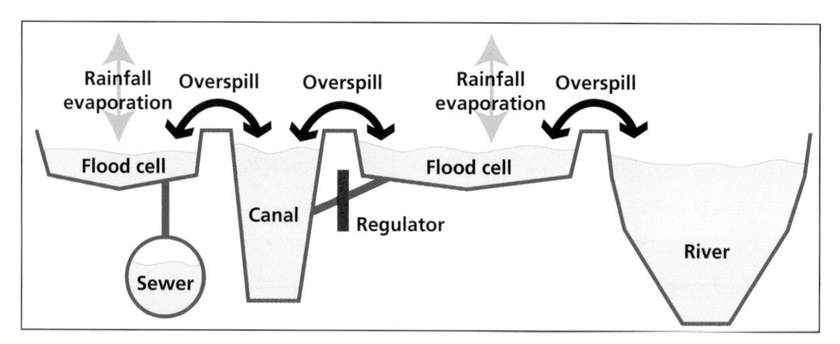

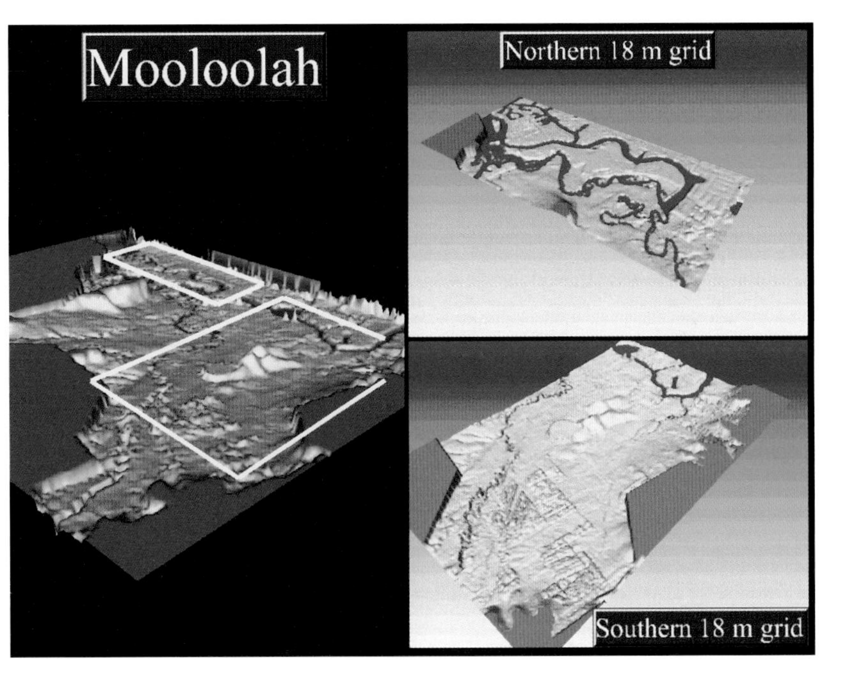

With the use of high-resolution digital elevation data, hydraulic and hydrodynamic models can simulate the temporal and spatial variation in flood levels in floodplains.

The second element of the software technology is a hydrodynamic model—a model that describes the network of rivers, canals and sewers in the floodplain and which can be set up semi-automatically from the GIS interface. Pumps and control structures such as weirs and culverts can also be included in the model. The flood-prone areas are represented as cells, and the overspill to and drainage from these areas is modeled. This model takes as its input the output from the hydrological model.

The output from the hydrodynamic model consists of information on the water level and the flow in the river reaches. A one-dimensional network model like MIKE 11 can be used to provide a pseudo-two-dimensional description in the floodplain areas. Coupled with detailed terrain data, it is possible with ArcView to obtain excellent visual representations of the floodplain. When that is done, floodwater levels can be viewed in combination with the local topography and the basic infrastructure. When elevation data is also brought into the picture, depth and flood-duration maps are made possible.

Flood basin management

A key element of the holistic approach to flood management that underlies technology is the protection and the restoration of natural forms of water storage. For example, forests are one of the densest biomass reserves on earth; they are able to store large volumes of water and return them to the atmosphere through evapotransportation. Trees also hold thin mountaintop soil in place, thereby reducing the problems that would otherwise be caused by sedimentation washing downstream. Wetlands—the areas between the land and the water—also play an important role in flood management because they can absorb both flood-water and sediment.

Forests alleviate flood hazards by storing water from rainfall. Wetlands provide another important means of storing floodwater. This decreases the risk of flooding in urban developments.

Forests and wetlands—natural flood inhibitors—are under threat and vast areas have already been destroyed. MIKE 11 can be used to assess the effect of regenerating forests and wetlands, or even creating new forms of natural storage. Moreover, by supplying the model with worst-case rainfall data, the effectiveness of a variety of flood-management plans can be assessed under extreme conditions.

Structural flood management

Although this approach to flood prevention is a departure from the traditional method of constructing dams, culverts and other physical structures, physical structures can still play a vital role in flood management. Even so, these structures are clearly different from the traditional ones, which tend only to migrate water from the flood-prone area. Newer structures emphasize water storage and include reservoirs in the catchment areas, as well as river channels that have increased capacity and retention basins in their lower reaches. Again, when used this way, the technology can also be used as a planning tool to compare the effectiveness of competing schemes.

The software can even be used to take account of flood damage economics. By relating flood levels and flood extent to the economic uses of affected lands, financial losses for property damage, loss of livestock and loss of income can be evaluated. MIKE 11 can thus clearly show

Reservoirs can reduce the risk of downstream flooding by providing flood-control storage.

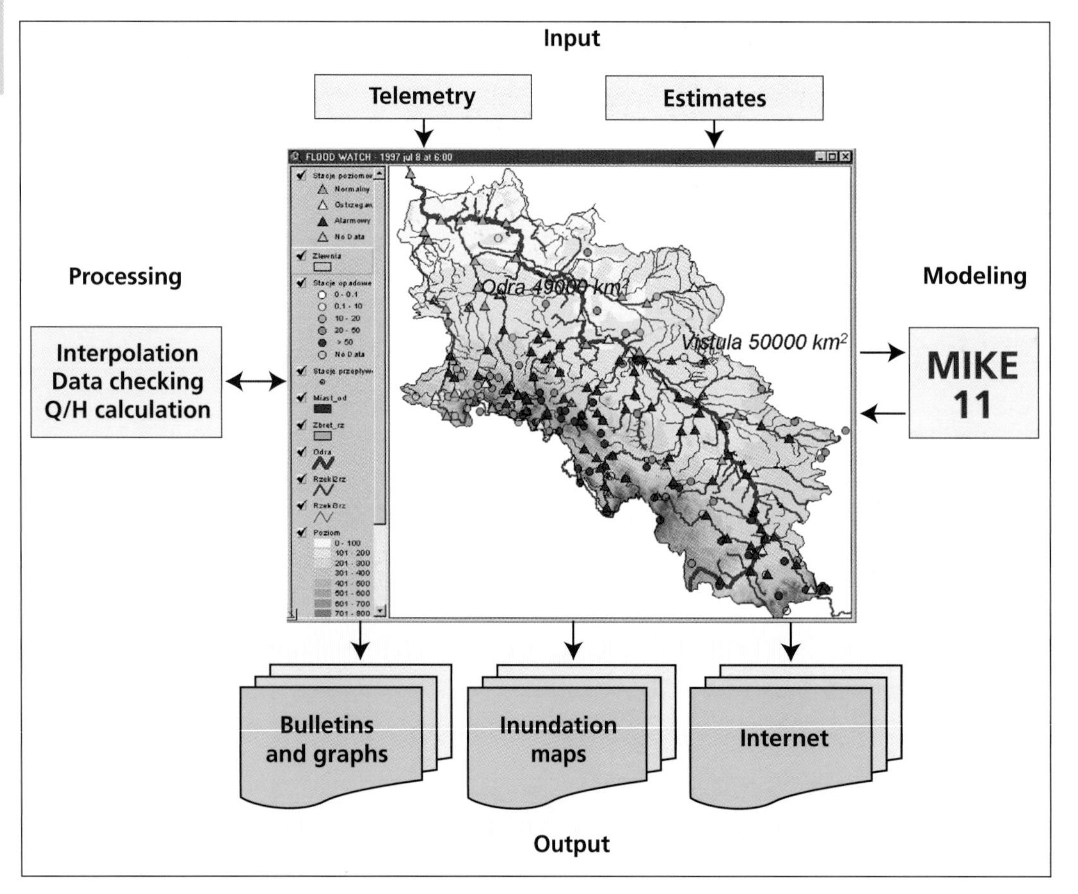

Real-time flood forecasting systems use hydraulic and hydrodynamic simulation plus links to telemetric networks. They provide an important tool for issuing flood warnings to communities in flood-prone areas.

basic cost comparisons—between taking no action and building reservoirs or retention basins.

Automated flood warning

For strategic planning, a MIKE 11 model can be used with historical or hypothetical rainfall data. But to provide more immediate flood warnings, the software must be fed with live data. The goal of any flood warning system is to provide warnings as early as possible, but the DHI technology has the advantage of not being dependent on the commencement of actual rainfall. Many of the Flood Watch systems take data from satellite images, from weather models and from local rainfall radar—all accessible before the start of actual precipitation.

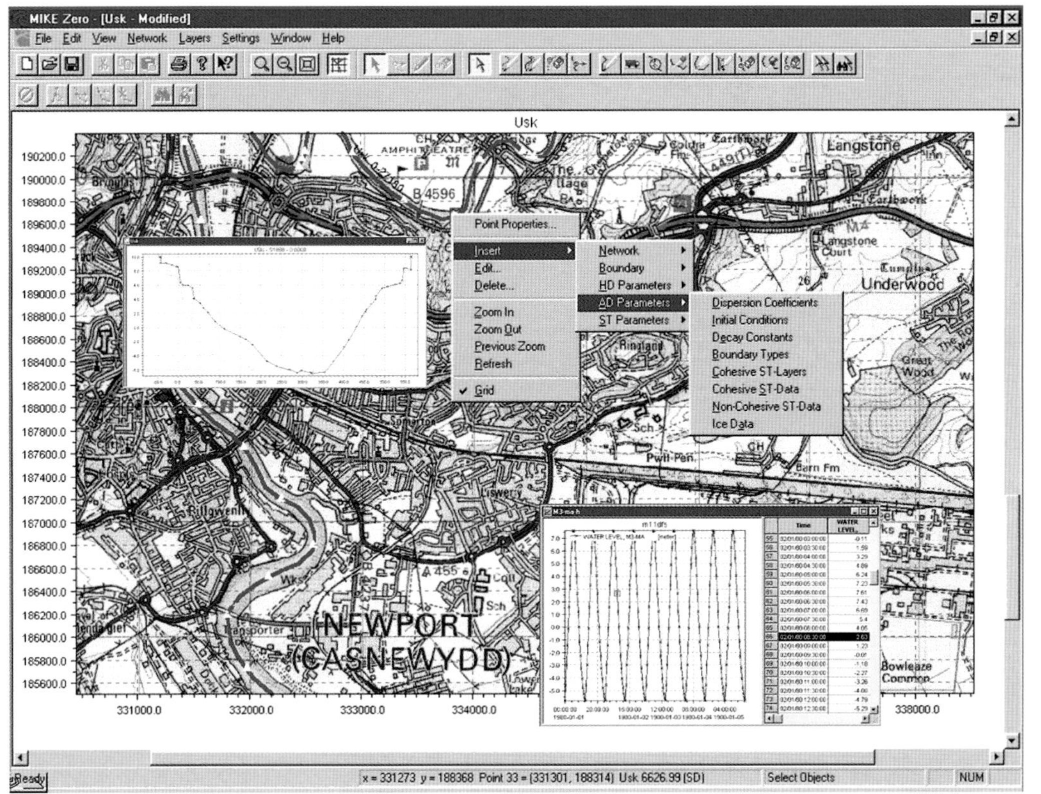

In this typical MIKE 11 interface, water levels of the River Usk in Wales are graphed.

Just as floods can be predicted with little user input, the Flood Watch system can also be used to disseminate warnings automatically. These warnings can take various forms: via telephone, for example, and through graphs and maps that can be faxed to interested organisations, including media outlets, so that radio and television stations can broadcast warnings to the public. The Web is also an important element of most Flood Watch systems. Web sites can present the information in various ways, including dynamic maps that show the extent of possible flooding.

Deluge swamps Italy

Due largely to the mountains that surround it on three sides, the Piedmont region of northern Italy has experienced severe flooding conditions every two years on average since record-keeping began more than two centuries ago. One of the worst floods on record occurred in October

2000; some called it a two-hundred-year event. Heavy and prolonged rainfall fell across the entire Po river basin and caused severe flooding, with water inundating vast areas and causing widespread damage and loss of life.

Fortuitously, a MIKE Flood Watch system had been installed for this area just weeks before. The severity of the event became clear: 32 000 people were evacuated from their homes, with 3000 of them rescued from isolated villages by helicopter. Estimates of the financial cost to Piedmont were in the billions of dollars. Twenty-nine people lost their lives, but the early warnings provided by MIKE software may have kept the toll from going much higher.

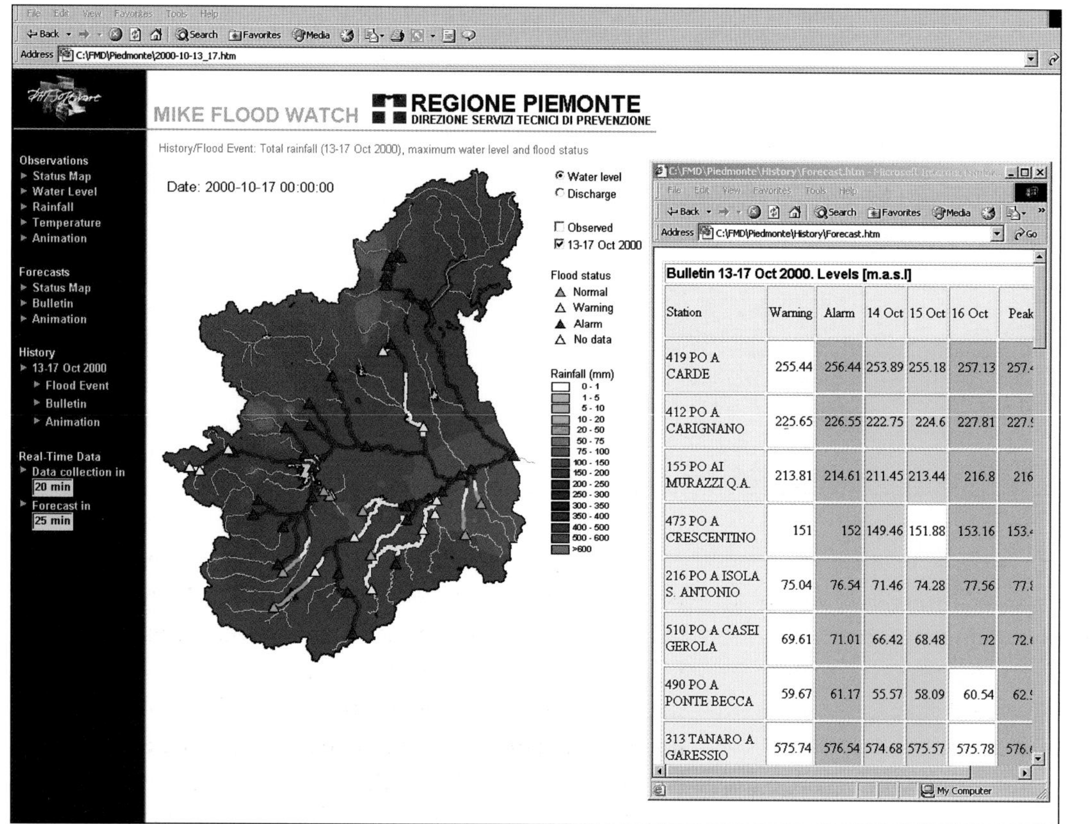

The Internet provides an efficient way to disseminate flood warning bulletins in Italy.

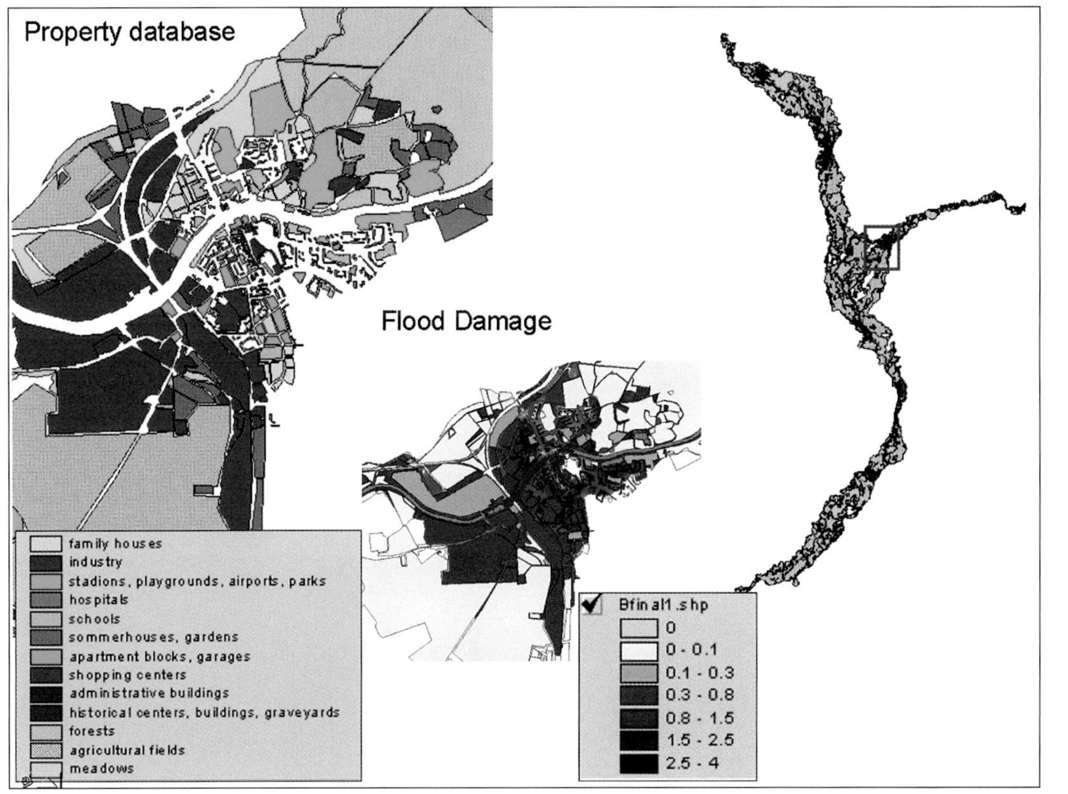

Flood maps, when combined with GIS information about the infrastructure, can provide information about the economic consequences of flooding.

Once bitten . . .

In 1997, the Czech Republic was devastated by floods when the River Morava burst its banks, causing an estimated $800 million (U.S.) in damage. Action was needed to prevent a recurrence. But the Czech authorities were keen to ensure that preventive action taken in one place wouldn't make matters worse somewhere else—a common problem with conventional flood-control measures.

Following requests from the Czech authorities, the Danish Environmental Protection Agency (DEPA) financed a transfer of Danish flood management technology, including MIKE 11, to the Czech Republic, which was used to assess the impact of various flood prevention schemes. Promising flood management options have been identified and simulation results have shown that these measures will significantly reduce flooding in the Morava catchment area.

Hardware

MIKE 11 uses dynamic memory allocation and the required DRAM varies depending on the actual setup and size of model domain. As a general guideline, a minimum of 128 MB DRAM and 1 GB of free disk space is recommended. Simulation times depend on the CPU; a 200-MHz Pentium II is recommended as the minimum.

Software

ArcView version 3.2, with the ArcView 3D Analyst extension.

Data

The main data requirements for MIKE 11 are detailed ground surface elevation data in the flood-prone areas; river cross-sections; and infrastructure such as bridges, weirs, dams, levees and houses. Data about spatial and temporal variation of rainfall in the catchment areas is also necessary. In connection with real-time flood forecasting and flood warning, MIKE 11 is linked directly to telemetric gaging stations and databases at the appropriate locations and meteorological offices.

Acknowledgments

Thanks to Börge Storm of DHI.

10 Safer Alpine communities

The phrase *water danger* conjures up raging floodwaters, rising perhaps to chimney level, a torrent sweeping villages away in its path. But, although we rarely consider it, water also exists as a solid on the earth's surface, and the danger posed by snow and ice is every bit as deadly to those living in mountainous areas as liquid floodwaters are to those living at lower elevations. Each winter, avalanches in the Alps and other mountainous regions around the world claim many lives and create destruction and tragedy. In this chapter we see how research being carried out at the Swiss Federal Institute for Snow and Avalanche Research (SLF) in Davos is being used to try to reduce the death and destruction caused by avalanches.

First tracks in the backcountry area

A slab avalanche released by off-piste skiers at a backcountry location in Davos

Snowfall brings a mixed blessing to the Alpine communities of Switzerland. Tourism accounts for one in twelve Swiss jobs and brings more than 20 billion Swiss francs ($15 billion U.S.) into the economy. With much of this tourist economy so focused on skiing and other winter sports, it's not surprising that there are strong pressures to further develop the Alpine settlements and infrastructure to create a bigger tourist economy and more jobs. But without careful consideration of avalanche risks, expanding the tourist facilities in Alpine valleys would almost certainly increase the risk of injury and damage caused by these natural disasters.

Artificially triggered powder snow avalanche at the SLF experimental site in Vallée de la Sionne

Snow pressure is threatening the roof construction of older buildings in Klosters.

However, the risk is not uniform—terrain conditions such as slope, aspect, curvature and confinement, as well as meteorological conditions, all affect the likelihood and severity of an avalanche. At one time, assessing the risk would involve rule-of-thumb and experience; while these methods have had their successes, neither is foolproof.

Today, mathematical modeling brings increased accuracy to determining avalanche risk. The results of this modeling process can be used to ensure that new developments are located in safer areas and that buildings in areas of moderate risk are built to withstand an avalanche. Furthermore, with the help of GIS, the clarity of avalanche forecasts has been improved, allowing people to take better precautions.

Historical perspective

In the last century, two major avalanche winters have struck the Swiss Alps. In the winter of 1950–51, about 1500 avalanches caused 98 deaths and damage to some 1500 buildings. In 1998–99, three heavy periods of snowfall within one month caused about 1200 large avalanche events, which in turn killed 17 people, caused significant damage to more than 1000 buildings, and seriously disrupted travel throughout the country.

The reduced number of fatalities and amount of damage in the 1999 event compared to the one a half-century earlier can be attributed, in part, to Swiss planning regulations introduced in the intervening years.

Among the changes was the creation of statutory avalanche hazard zones, designated according to the frequency of occurrence and impact force. Building is prohibited in red zones (the highest risk), whereas in blue zones (the next highest), buildings must be constructed to withstand a force of three tonnes per square metre. No restrictions apply to yellow zones, where flow avalanches are rare and only powder snow avalanches—those without a dense core at their base—occur. White zones represent areas that are supposed to be safe with respect to any type of avalanches.

The regulations were effective, but even 17 deaths in 1999 led many experts to believe that the zoning process had to be revised to more accurately evaluate the hazards of both powder snow and of multiple avalanche events. To overcome the perceived deficiencies in the regulations, new numerical simulation methods, in combination with GIS technology, have been applied to the process.

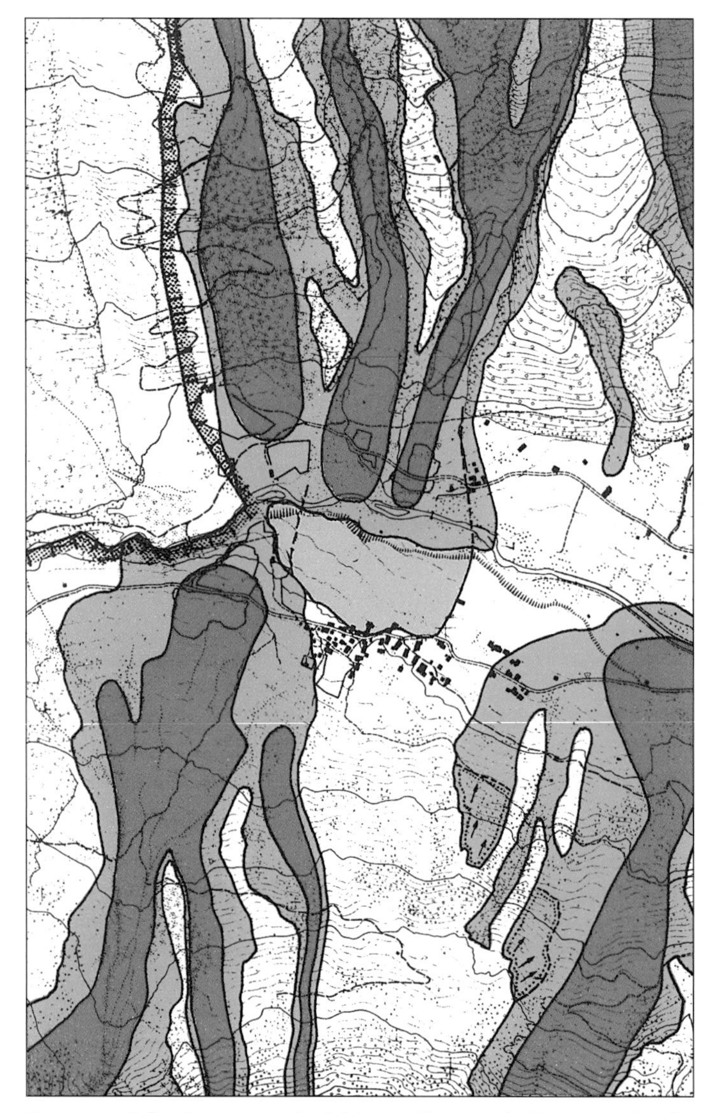

Example of the four zones (red, blue, yellow, white) in an avalanche hazard map of Switzerland

Avalanche damage to a building in Valzur, Tirol, Austria in 1999

A completely filled snow net above Davos. Such structures proved to be very effective in 1999. No avalanches occurred on slopes equipped with defence structures.

New buildings in hazard areas must be constructed to withstand a force of three tonnes per square metre.

Of course, even if planning regulations and perfect zoning were to become a reality, loss of life and damage to property could never be eliminated entirely. Older buildings may still be in hazardous areas and skiers would still be at risk in the event of an avalanche. Therefore, a second initiative has been created: the development and continuing improvement of an early warning system.

These two initiatives are giving researchers hope and expectation that loss of life and disruption of the local economy will be reduced further when high-avalanche-danger conditions exist in the Swiss Alps.

Image © swisstopo

Deposits of three large consecutive avalanche events in February 1999 in Geschinen, Switzerland

Hazard mapping

Traditionally, avalanche hazard maps in Switzerland have been based on calculations of avalanche dynamics that use the so-called Voellmy-Salm mathematical model. This hydraulic model contains only two friction parameters, one

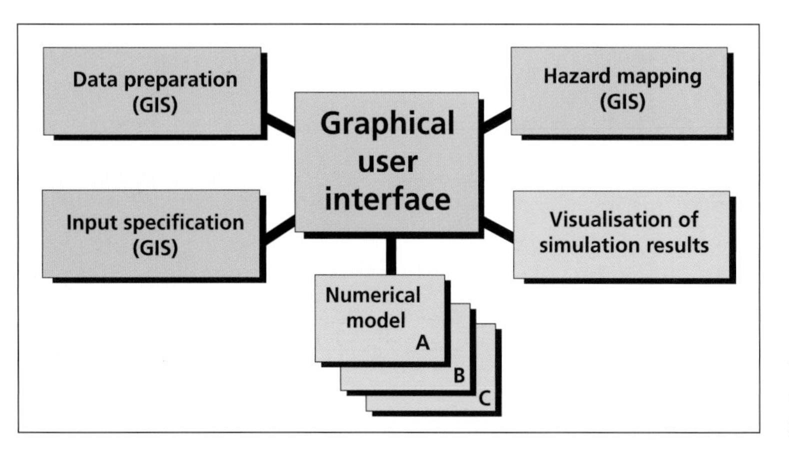

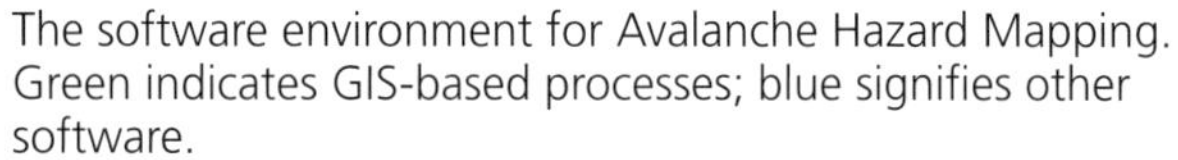

The software environment for Avalanche Hazard Mapping. Green indicates GIS-based processes; blue signifies other software.

for dry friction and the other for turbulent or viscous friction. During the 1990s, one- and two-dimensional numerical depth-averaged continuum models were developed that resolved many of the shortcomings of the Voellmy-Salm model. However, numerical models also have their drawbacks. They require a more detailed description of the avalanche release zone, and they generate large amounts of output data that are difficult to analyse. To overcome these problems, the numerical models were integrated into a GIS-based user environment.

The diagram on the preceding page shows how an ArcInfo graphical user interface (GUI) connects all the different elements of the avalanche dynamics calculation. The GUI incorporates other modules, including those for topographical data preparation, input specification and hazard mapping. The numerical simulation models and their visualisation tools are implemented using other software but are accessed from the ArcInfo GUI.

Terrain analysis and input specification

A good digital representation of the topography is essential input to the model. Currently, a digital elevation model (DEM) of avalanche areas is available only in raster format with a spatial resolution of 25 metres. This is an adequate resolution for open slope terrain but it's not accurate enough for the steep gullies that form part of

Avalanche release areas such as this one in Geschinen, Switzerland, are often quite complex. GIS and DEM are used to find rules to assess potential release areas for extreme avalanche events.

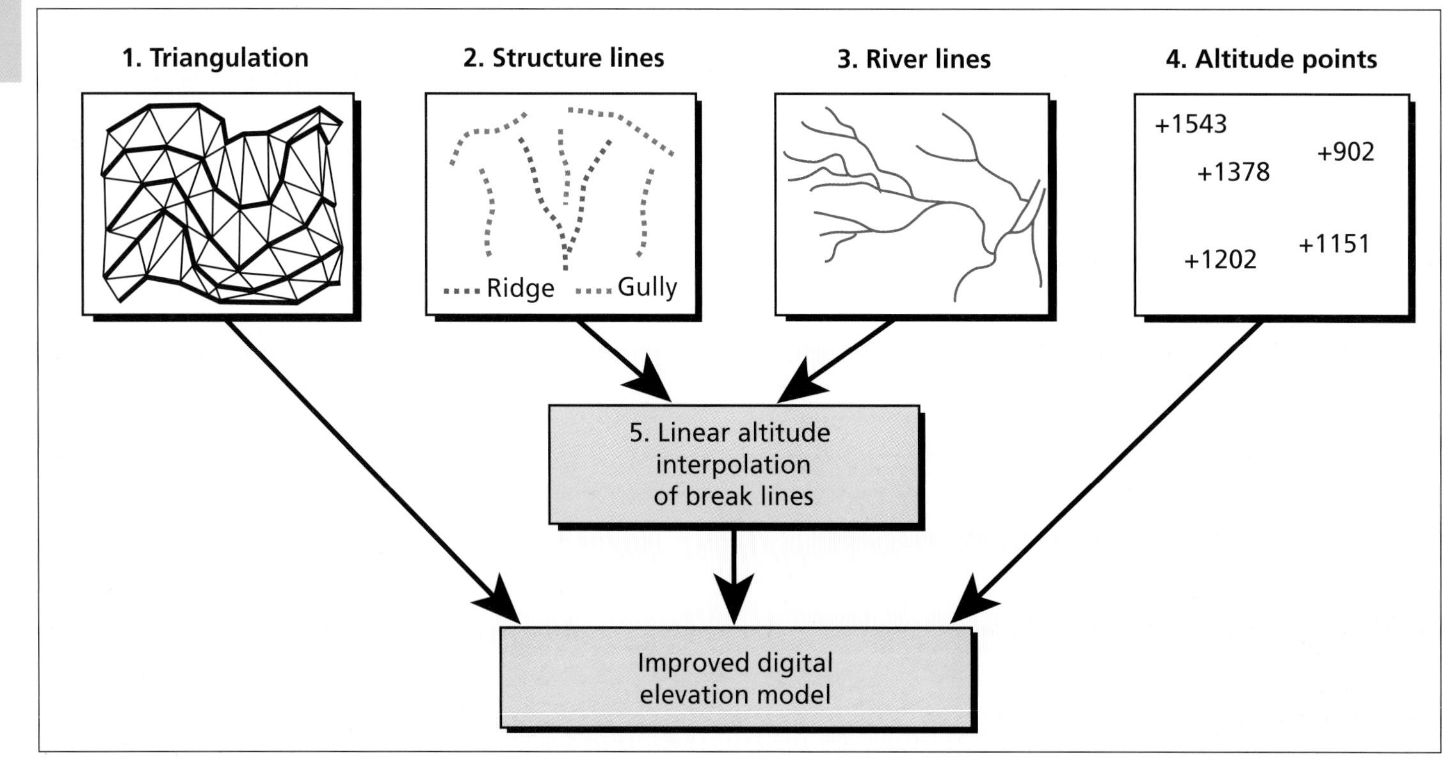

Flowchart of the improvement of the DEM based on the integration of contour lines and various break lines, with and without z-values and altitude points.

most avalanche paths. Additional accuracy in these areas is provided by the data preparation module that integrates contour lines, various break lines—with and without altitude values—and altitude points, using standard ArcInfo procedures. In addition, a linear altitude interpolation tool is used for creating break lines with missing altitude values.

In addition to this topographic information, the mathematical model needs various other parameters in order to calculate the avalanche run-out distance and velocities. This essential input data includes a definition of the snow release area and estimate of the snow depth, as well as friction and entrainment parameters along the avalanche path. The input specification module allows this input data to be entered directly onto a digitised map.

Map production

The numerical model results provide information about flow and deposition depths, velocities and impact pressures. The impact pressure is the most important result for avalanche hazard mapping since this is the figure by which the statutory zones are defined. The GIS displays these pressure forces on maps and also shows the results of different scenarios (such as return periods or different release area assumptions) on a combined pressure zone map according to Swiss guidelines. For the future there are plans to improve the visualisation of the numerical results by creating 3-D perspective views of avalanche tracks with draped orthophotos or maps as background, and to then animate the avalanche simulations.

The computer models can also be used to simulate the interaction of avalanches with dams. Two such interactions in the Obergoms Valley, below, and in Iceland, left, are shown. Flow depth of the avalanching snow is in metres. The simulations help determine the dimensions and effects that a preventive deflection measure such as a dam provides.

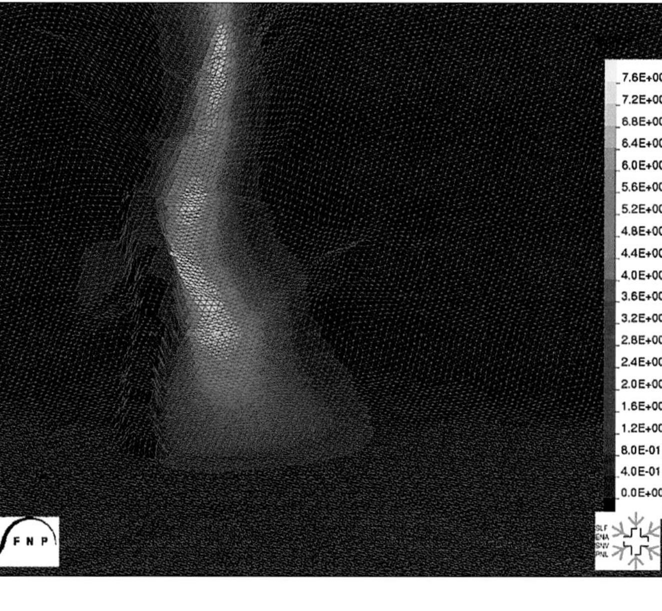

Avalanche warning system

The Swiss avalanche warning system has undergone a major development in the last few years, making extensive use of modern developments in information and communication technology. A network of automatic stations for the measurement of meteorological and snow parameters has been built. This data, together with that from a dense network of human observers, form the basis for the daily avalanche forecasting for the Swiss Alps issued by SLF.

Avalanche forecasters at SLF use a wide range of tools to analyse the data and to prepare warning bulletins. Although avalanche forecasting is still a manual job requiring extensive skill and experience, the use of statistical analysis tools and GIS has improved productivity and streamlined the dissemination of the forecasts. Quick and easy access to the information is crucial for skiers—so

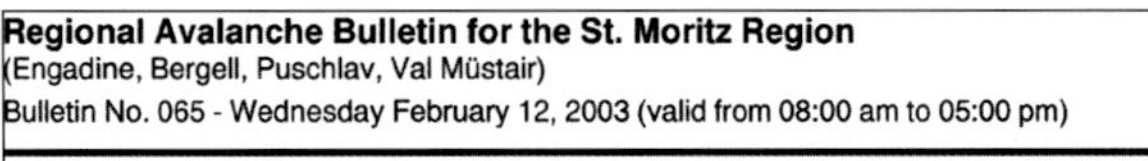

Regional Avalanche Bulletin for the St. Moritz Region
(Engadine, Bergell, Puschlav, Val Müstair)
Bulletin No. 065 - Wednesday February 12, 2003 (valid from 08:00 am to 05:00 pm)

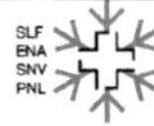

In some areas still considerable hazard of avalanches

The structure of the snow cover is variable and locally weak. Dangerous areas are in gullies and bowls and particularly in shady slopes. The transition from shallow to deep snowpacks should be traversed with long safety distances. Gliding snow avalanches can occur below roughly 2400 m a.s.l..

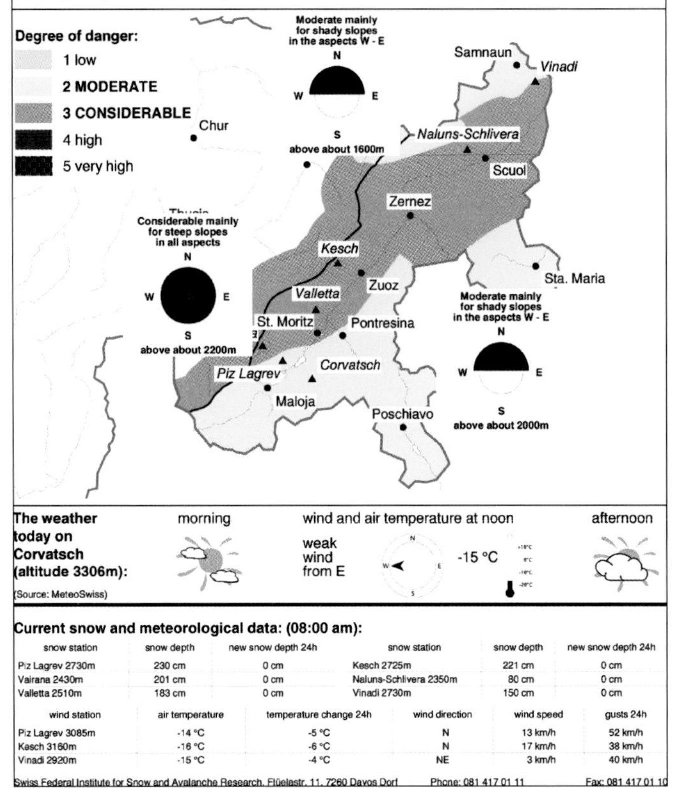

Current snow and meteorological data: (08:00 am):

snow station	snow depth	new snow depth 24h	snow station	snow depth	new snow depth 24h
Piz Lagrev 2730m	230 cm	0 cm	Kesch 2725m	221 cm	0 cm
Vairana 2430m	201 cm	0 cm	Naluns-Schlivera 2350m	80 cm	0 cm
Valletta 2510m	183 cm	0 cm	Vinadi 2730m	150 cm	0 cm

wind station	air temperature	temperature change 24h	wind direction	wind speed	gusts 24h
Piz Lagrev 3085m	-14 °C	-5 °C	N	13 km/h	52 km/h
Kesch 3160m	-16 °C	-6 °C	N	17 km/h	38 km/h
Vinadi 2920m	-15 °C	-4 °C	NE	3 km/h	40 km/h

Swiss Federal Institute for Snow and Avalanche Research, Flüelastr. 11, 7260 Davos Dorf Phone: 081 417 01 11 Fax: 081 417 01 10

Example of an avalanche forecast bulletin for the St. Moritz region. Morning bulletins are dominated by a graphical display of the spatial distribution of danger levels. The regional resolution and use of graphics helps readability and interpretation.

Automatic station for meteorological and snow parameter measurements

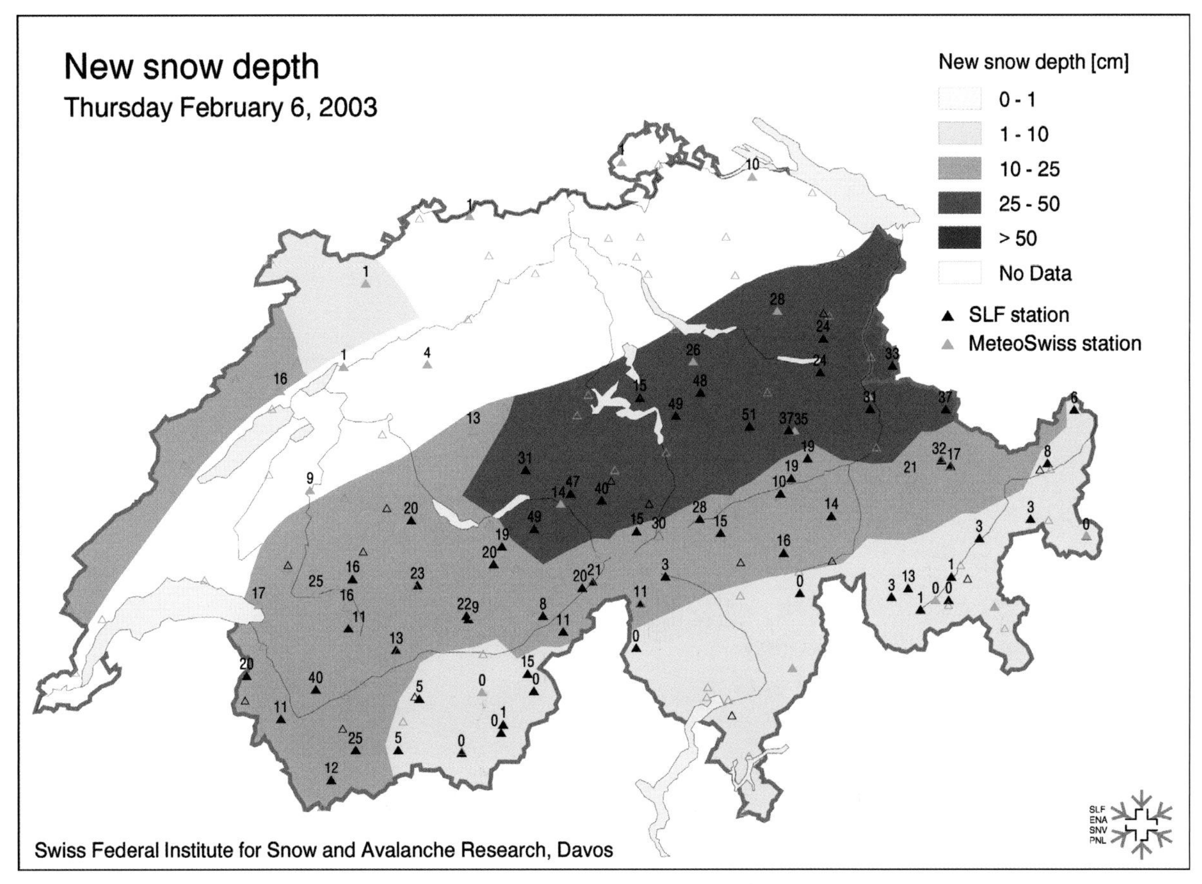

Maps of Switzerland like this one, showing new snow depth, are created from an Inverse Distance Weighting (IDW) interpolation function in the GIS, using manual measurements made at weather stations, indicated by the green and black triangles. The snow depth information is incorporated into the SLF avalanche warning system and also disseminated to the public.

the information is disseminated through a wide array of media, including dedicated phone numbers, fax, Internet, SMS, WAP, and as a print product available at hotels and ski-lift waiting areas. On the Web, a text-based bulletin is accompanied by a series of GIS-based maps and includes additional information about danger levels, snow depth, new snow depth and snow cover stability.

Avalanche forecasters at SLF analysing maps and statistical output to prepare warning bulletins.

Besides the publicly available information, the security services responsible for traffic, buildings and other infrastructure have access to additional information. An important support tool for officials of those services is the early warning system. In case of an imminent critical situation, an early warning message is released that indicates which regions are most at risk.

The period of serious avalanches in 1999 was a major test of the system and the people and organisations involved in crisis management. Among the lessons learned was that, in

Avalanche forecaster at SLF editing a danger-level map

addition to good information flow from the early warning system to the security services, a good information flow was crucial for the parties involved in crisis management. As a consequence, SLF, together with the Alpine Security and Information Center (ASI) in Austria, have developed an Internet-based Management Information System, IFKIS-MIS. This allows security officials to communicate with each other about avalanches and the actions they have taken. Future plans include the development of an Internet map server for display, query and analysis of all spatial information.

In the past, building new ski resorts—and thereby helping the Swiss economy—had the unfortunate side effect of putting greater numbers of people at risk from the ever-present threat of avalanches. New planning regulations and early warning systems, both of which rely on GIS-based avalanche modeling, have changed that equation. Although danger can never be eliminated entirely, the Alpine ski resorts are now safer than ever before for Switzerland's ever-growing number of visitors intent on enjoying fun on the slopes.

Hardware

Application and database servers: Sun Enterprise™ 250 (Solaris™ 8)

Workstations: Sun Blade™ 100 (Solaris 8), Compaq desktop PCs (Windows 2000)

Software

ArcInfo 8.2, ArcView 3.2

Interactive Data Language 5.4

C/C++

Data

Snow, weather and danger-level assessment databases in Oracle 8*i*; countrywide snow and weather data from automatic and manual measurement stations; daily historic avalanche danger-level assessments from every region; ArcInfo avalanche coverages; digital avalanche data and historic records from the Davos, Bernese Oberland and Zuoz regions

Web site

www.slf.ch

Acknowledgments

Thanks to Andreas Stoffel and Urs Gruber of the Swiss Federal Institute for Snow and Avalanche Research (SLF).

11

De Maaswerken: Making space for several solutions

Anyone who lives near a river knows there is always a chance that river could overflow its banks and inundate the surrounding neighbourhoods. But when the passage of time brings no such problems, complacency can often overtake even the most vigilant homeowner.

High water is an ever-present reality in the lives of the residents of the catchment area of the River Meuse.

The Web site of The Meuse Works provides a centralised source of information for residents who could be affected by the construction activity.

Residents of the catchment area of the River Meuse, in the southern Dutch province of Limburg, know this situation only too well. In 1993, after a long period without serious problems, the water in the Meuse rose to levels that flooded several villages and forced the evacuation of thousands of people. When the situation repeated itself in 1995, the government decided that steps had to be taken to prevent a repetition of the devastation.

These new measures meant a radical transformation in the catchment area, where flooding is predicted to occur every 50 years. In fact, even before the 1993 floods, teams of engineers had been working out just which measures might need to be taken and where they might be implemented. Operating with a renewed urgency, the diverse initiatives were brought together under one umbrella project called De Maaswerken (The Meuse Works). Under De Maaswerken, various government organisations have been collaborating to find the best solutions to reduce flooding, with the goal of protecting families from floods up to and including 250-year events.

GIS specialists have worked closely with hydrologic experts to provide information to the Maaswerken decision makers, who must choose among a wide variety of flood-protection measures.

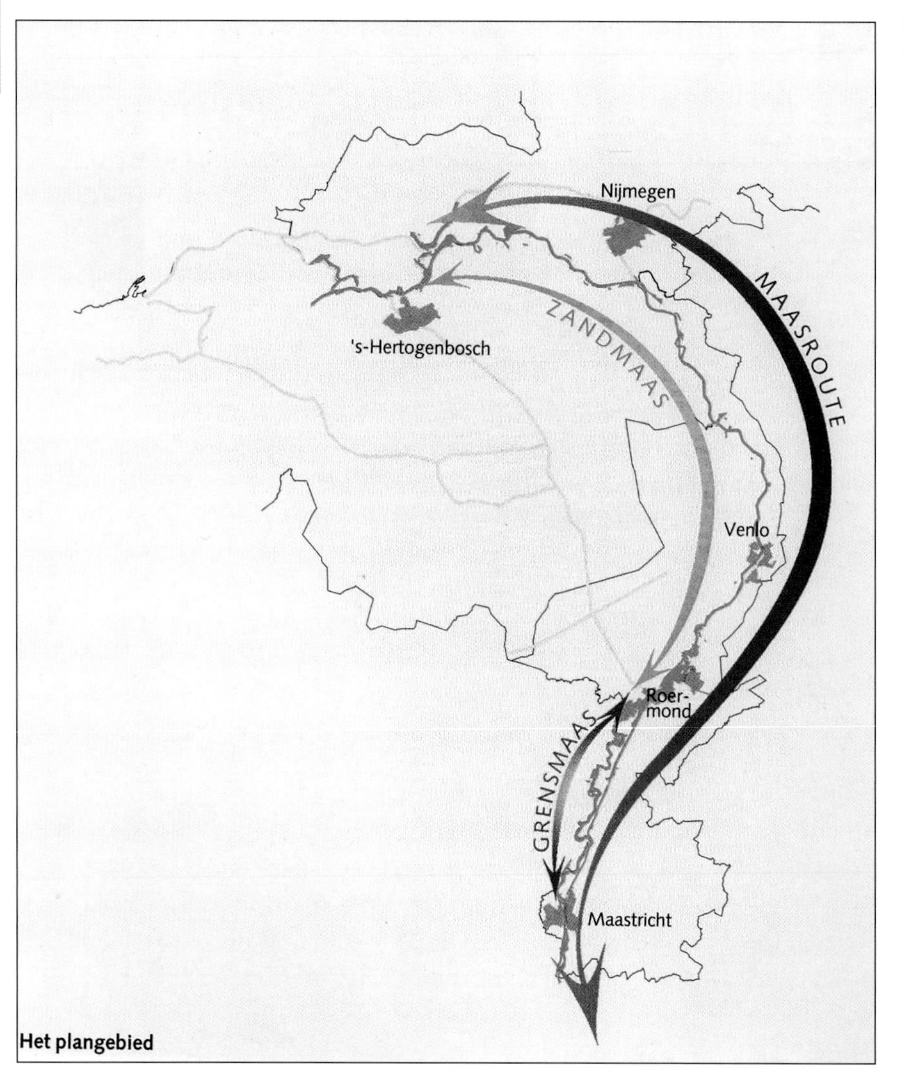

The larger Meuse Works project is divided into three smaller projects with different but related objectives.

Maps are a critical part of The Meuse Works project, showing not only what work is planned, but also whether financing has been put in place. Here, projects such as locks and dams, and widening and deepening of the river, for which funds have been found, are shown in green; projects in red are planned but still awaiting funding.

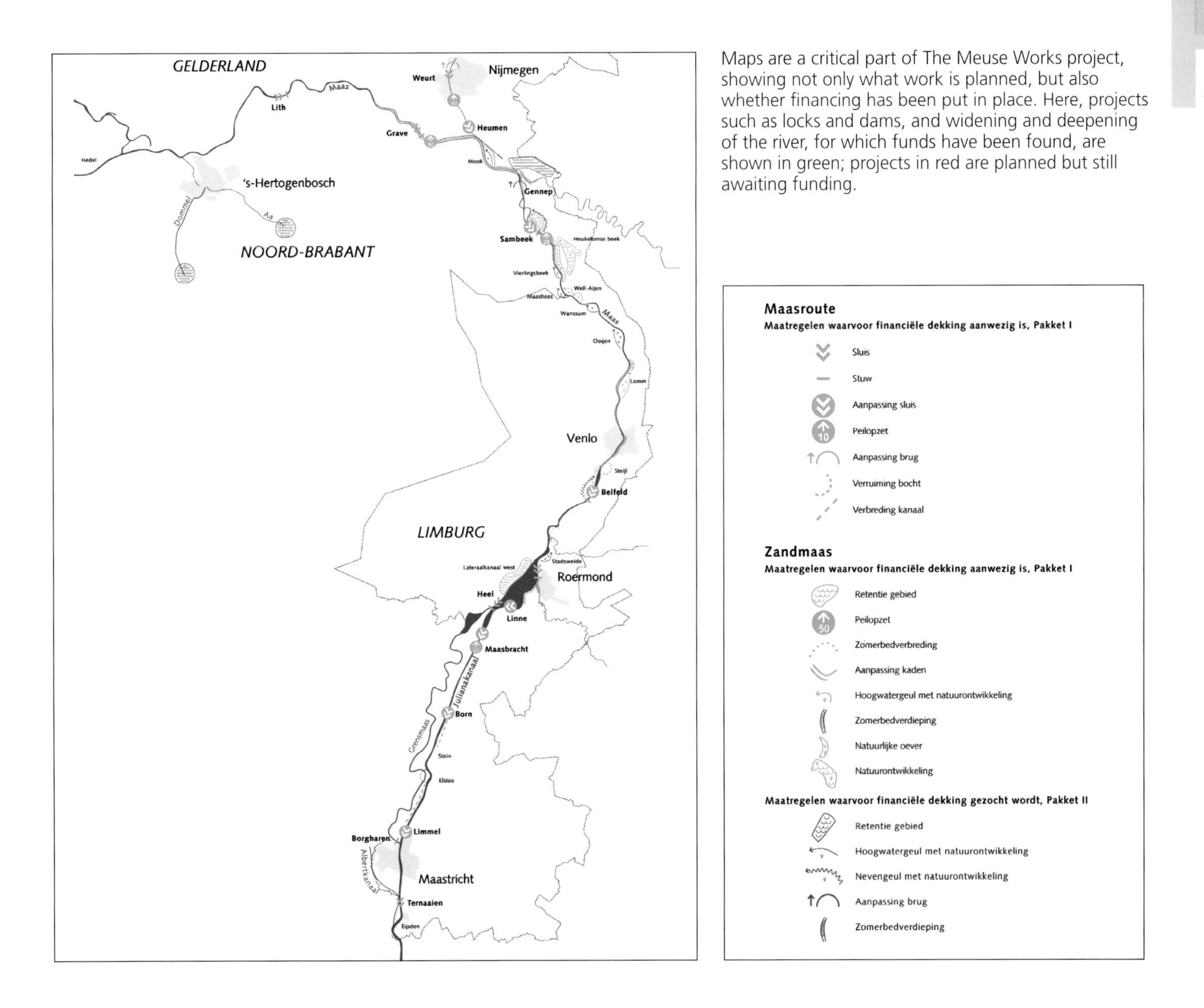

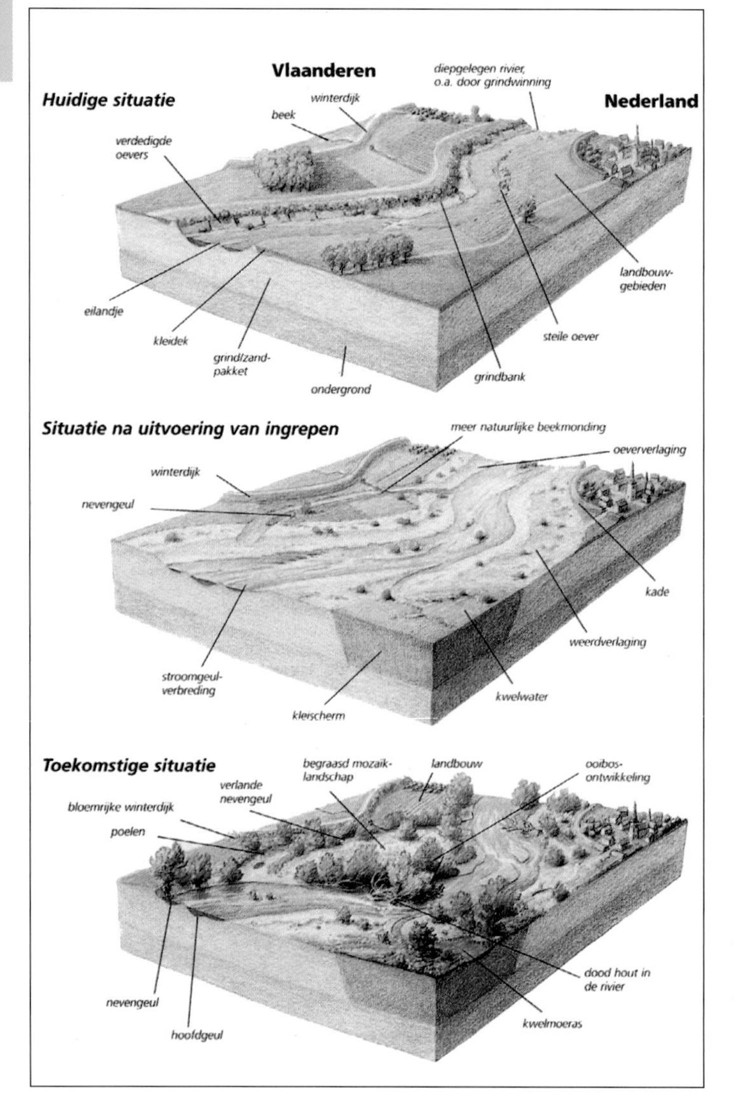

Three-dimensional models show the potential effects of the project on one section of the river near the Belgian border. The top model shows the current situation, the middle model shows what the immediate effects will be on the surrounding countryside, and the bottom model shows how the area will look after vegetation has matured and the environment has adapted to the changes to the river.

Three projects

The Maaswerken project is divided into three smaller ones, called Grensmaas, Zandmaas and Maasroute.

The Grensmaas project covers the first stretch of the Meuse in the Netherlands, near the border with Belgium. This portion of the river twists and meanders and is unnavigable when the water level is low. Barge traffic must use the parallel Juliana canal rather than the river. One notable feature of the river in this section is the unique gravel of the riverbed, which is found nowhere else in the Netherlands.

After Linne, 45 kilometres north of Maastricht, the river continues for 148 kilometres, until 's-Hertogenbosch. Here the river is known as the Zandmaas. In the Zandmaas project, flood protection efforts are combined with development of small-scale natural habitats.

The Maasroute project encompasses both the Zandmaas section and its related canals and is designed to improve the navigation channel of both components.

The challenge: Integrated planning

The new plans for the Meuse illustrate the significant shift in flood prevention policy in the Netherlands during the past 15 years. Traditionally, river flooding in an area mandated merely the raising of old dykes or construction of new ones. Now, flood control engineering techniques have focused more on creating additional space for rivers by widening them and deepening the riverbed. Only in areas where it is impossible to widen the river does the building or raising of dykes become a consideration.

And, although the main challenge for the government organisations was the safety of citizens living in the catchment area of the Meuse, there was a second, equally important challenge—to integrate the new flood protection measures with all of the existing plans for the Meuse and surrounding areas. Several synergies resulted from this integrated approach.

Advantages of integrated planning

The Grensmaas project, for example, combines several interrelated efforts: flood protection, gravel extraction and the development of natural habitats. Profits from the extraction of sand and gravel, which is needed to give the river more space, will in turn help finance the new habitat areas.

In the Zandmaas project, riverbanks in portions of the river where there are bends and turns will be widened to reduce the chances of debris clogging the river channel when water levels are high. Widening the river in areas where it is particularly meandering will also make it easier for barges to navigate the river, which is the objective of the Maasroute project.

Dredging of the Meuse will make it more navigable.

Organisational advantages of an integrated approach have included establishing a central administration office for land acquisition. The Meuse Valley Land Office is part of The Meuse Works project and is charged with acquiring land by mutual agreement with landowners.

The public also reaps additional benefits from an integrated approach. During public discussions of the projects, residents were presented not with an assortment of fragmented projects, without apparent connection, but rather a plan that integrates and connects many aspects of their local geographies.

From data to maps

Not surprisingly, as part of the overall project design, a great deal of research has been done into flood protection measures and their effects. For example, project designers created four alternate sets of measures, each with a different emphasis with regard to the development of natural habitats, to safety and to costs. The effects of these measures in terms of rates of flow, water levels, the level of the riverbed and frictional resistance in the ground were calculated with detailed hydraulic models such as WAQUA.

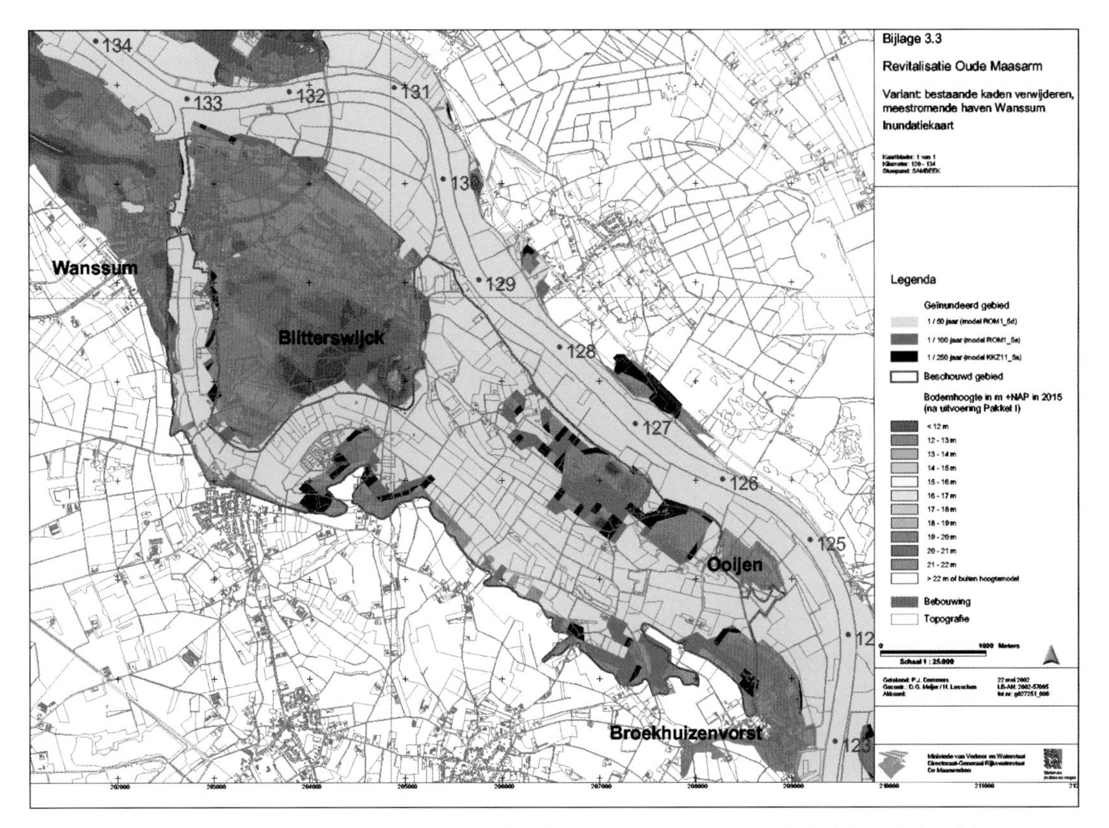

This flood-prediction map shows areas likely to flood once every 50 years in light blue; darker blues indicate areas at risk every 100 and 200 years. Red numbers indicate river length within the Netherlands.

The GIS task force produces programmes that convert hydraulic calculations from WAQUA automatically into GIS maps. These GIS tools were helpful in assessing the potential consequences of the various flood protection measures. The data from the model also incorporated specific conditions on the ground, such as the height of river quays. This made it possible to show on a map which areas are likely to inundate—with specific high-water discharge in cubic metres per second—before and after flood protection measures are implemented.

A central digital library containing geographic data on the area covered by the project formed an important basis for the multifaceted applications of GIS during the whole project.

Ecotopen Generator

Chief among the notable environmental tools developed with GIS is the Ecotopen Generator, created with ArcInfo. This application shows the locations and varieties of ecotopes that could develop in an area, given specific conditions such as the level of the riverbed and the velocity and depth of water.

The Ecotopen Generator, using grids, calculates in which direction an ecotope, such as woodland, will spread once it has started to develop and nature is left to itself.

The software designers establish the rules. They set out maximum or minimum speeds of flow that will allow scrub or woodland to develop, or maximum percentages of specific habitats that can be permitted on the riverbank.

It is an iterative process; the outcome of one analysis forms the input to the next. Each ecotope has a certain resistance with regard to the flow of water. This is used as input for the hydraulic models. The output of the hydraulic calculations is the input for predicting changes in the development of ecotopes, which in turn leads to a different pattern of resistance.

Planners can then take these calculations into account to balance safety measures with environmental considerations, to determine how far nature can be allowed to run its course without impeding the effectiveness of the flood protection measures.

The Ecotopen Generator works so well that it earned its inventor his doctoral degree.

GIS for public information

GIS was also used in the public consultation process. Although flood protection measures are needed to protect the population, it is also true, paradoxically, that those measures cannot be implemented without some adverse effects on the same population. Such measures as rerouting of dykes and excavation work near villages affect the immediate, day-to-day environment of local residents. Through public consultation, which is required by law anyway, interested parties such as local residents were able to react to the plans and to register any objection. Maps relevant to the proposals were published on the Internet, using ArcIMS technology, which could handle storage requirements for both the online maps and the underlying data. The maps consist of several layers. A visitor to the project's Internet site can choose which information she wants to see, examine the whole plan area, or zoom in and out to find out specific details about a proposed measure.

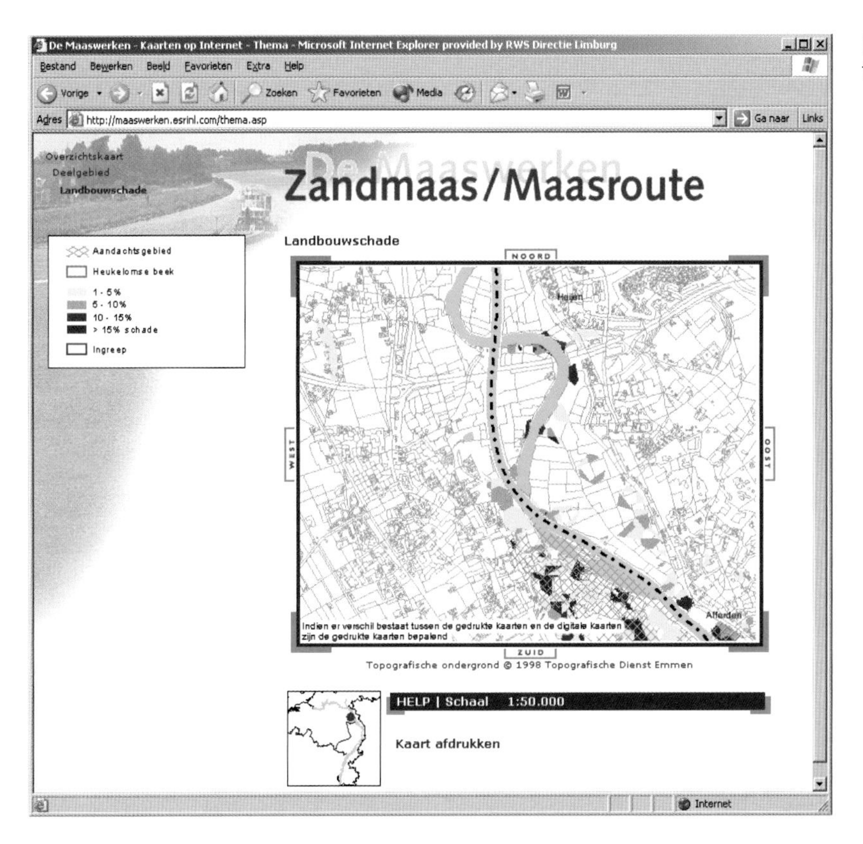

Dark red areas show the damage to agricultural areas as the result of the revitalisation of the Meuse.

Last steps

Although the project planning is nearing its end, GIS technology is still aiding the process. Administrators are using GIS for land acquisition and for communication with contractors on such day-to-day issues as which roads will be used by lorries transporting soil.

Software
AutoCAD
ArcIMS
ArcView
ArcInfo
Oracle

Acknowledgments
Thanks to Rijkswaterstaat Maaswerken, a joint venture of the Ministry of Transport, Public Works and Water Management; the Province of Limburg; and the Ministry of Agriculture, Nature Management and Food Quality.

Books from ESRI Press

ISBN	Title
1-58948-052-X	A System for Survival: GIS and Sustainable Development
1-58948-073-2	Advanced Spatial Analysis: The CASA Book of GIS
1-58948-034-1	Arc Hydro: GIS for Water Resources
1-58948-074-0	ArcGIS and the Digital City: A Hands-on Approach for Local Government
1-879102-51-X	ArcView® GIS Means Business
1-879102-79-X	Beyond Maps: GIS and Decision Making in Local Government
1-58948-044-9	Cartographica Extraordinaire: The Historical Map Transformed
1-58948-023-6	Community Geography: GIS in Action
1-58948-051-1	Community Geography: GIS in Action Teacher's Guide
1-58948-040-6	Confronting Catastrophe: A GIS Handbook
1-58948-075-9	Connecting Our World: GIS Web Services
1-58948-024-4	Conservation Geography: Case Studies in GIS, Computer Mapping, and Activism
1-58948-021-X	Designing Geodatabases: Case Studies in GIS Data Modeling
1-879102-88-9	Disaster Response: GIS for Public Safety
1-879102-48-X	Enterprise GIS for Energy Companies
1-879102-05-6	Extending ArcView GIS (version 3.x)
1-58948-083-X	Getting to Know ArcGIS® Desktop: Basics of ArcView, ArcEditor™, and ArcInfo®, Second Edition, Updated for ArcGIS 9
1-58948-018-X	Getting to Know ArcObjects™: Programming ArcGIS with VBA
1-879102-46-3	Getting to Know ArcView GIS (version 3.x)
1-58948-077-5	GIS and Land Records: The ArcGIS Parcel Data Model
1-58948-056-2	GIS for Everyone, Third Edition
1-879102-65-X	GIS for Health Organizations
1-879102-64-1	GIS for Landscape Architects
1-879102-66-8	GIS in Public Policy: Using Geographic Information for More Effective Government
1-879102-85-4	GIS in Schools
1-879102-86-2	GIS in Telecommunications

Continued on next page

Books from ESRI Press (continued)

ISBN	Title
1-58948-033-3	GIS Means Business, Volume II
1-879102-80-3	Hydrologic and Hydraulic Modeling Support with Geographic Information Systems
1-879102-81-1	Integrating GIS and the Global Positioning System
1-58948-071-6	Making Community Connections: The Orton Family Foundation Community Mapping Program
1-879102-53-6	Managing Natural Resources with GIS
1-58948-014-7	Mapping Census 2000: The Geography of U.S. Diversity
1-58948-022-8	Mapping Our World: GIS Lessons for Educators, ArcView 3.x Edition
1-58948-080-5	Mapping the Future of America's National Parks: Stewardship through Geographic Information Systems
1-58948-072-4	Mapping the News: Case Studies in GIS and Journalism
1-58948-045-7	Marine Geography: GIS for the Oceans and Seas
1-58948-088-0	Measuring Up: The Business Case for GIS
1-879102-62-5	Modeling Our World: The ESRI® Guide to Geodatabase Design
1-58948-032-5	Past Time, Past Place: GIS for History
1-58948-011-2	Planning Support Systems: Integrating Geographic Information Systems, Models, and Visualization Tools
1-58948-043-0	Salton Sea Atlas
1-879102-06-4	The ESRI Guide to GIS Analysis, Volume 1: Geographic Patterns and Relationships
1-879102-78-1	The ESRI Press Dictionary of GIS Terminology
1-58948-070-8	Thinking About GIS: Geographic Information System Planning for Managers
1-879102-47-1	Transportation GIS
1-58948-016-3	Undersea with GIS
1-58948-113-5	Unlocking the Census with GIS
1-879102-50-1	Zeroing In: Geographic Information Systems at Work in the Community

ESRI Press publishes a growing list of GIS-related books. Ask for them at your local bookstore or shop for them online at your favorite Web retailer.

ESRI Press • 380 New York Street • Redlands, California 92373-8100 • www.esri.com/esripress

GIS for Water Management in Europe
Book design, production, and image editing by Savitri Brant
Cover design by Amaree Israngkura
Copyediting by Michael J. Hyatt
Printing coordination by Cliff Crabbe